A Preliminary Study of the Structural Dynamic Behavior of the NASA Manned Spacecraft Center (MSC) Centrifuge

by

Frederick W. Palmieri

ISBN: 1-58112-206-3

DISSERTATION.COM

USA   2003

A Preliminary Study of the Structural Dynamic Behavior
of the NASA Manned Spacecraft Center (MSC) Centrifuge

Dissertation.com
USA   2003

ISBN: 1-58112-206-3
www.Dissertation.com/library/1122063a.htm

A PRELIMINARY STUDY OF THE STRUCTURAL DYNAMIC BEHAVIOR OF THE NASA MANNED SPACECRAFT CENTER (MSC) CENTRIFUGE

A DISSERTATION

SUBMITTED TO THE DEPARTMENT OF CIVIL ENGINEERING

OF MADISON UNIVERSITY

IN PARTIAL FULFILLMENT OF THE REQUIREMENTS

FOR THE DEGREE OF

DOCTOR OF PHILOSOPHY

By

Frederick William Palmieri

August 22, 2003

I certify that I have read this thesis and that in my opinion it is fully adequate, in scope and in quality, as a dissertation for the degree of Doctor Of Philosophy.

I certify that I have read this thesis and that in my opinion it is fully adequate, in scope and in quality, as a dissertation for the degree of Doctor Of Philosophy.

I certify that I have read this thesis and that in my opinion it is fully adequate, in scope and in quality, as a dissertation for the degree of Doctor Of Philosophy.

Approved by the Graduate Committee of the University.

Abstract

This dissertation involves a preliminary study into the structural dynamic behavior of the NASA Manned Spacecraft Center (MSC), located in the Flight Acceleration Facility, bldg 29, in Houston, Texas. The 50-ft. arm can swing the three-man gondola to create g-forces astronauts will experience during controlled flight and during reentry. The centrifuge was designed primarily for training Apollo astronauts. During operation of the centrifuge, the astronauts can control the motion of the gondola in two gimbal axes, while the gondola is rotating about its principal axis, to simulate flight activity. The result of these coupled motions lead to transient loading functions, which arise due to rigid body kinematics.

The study is describe in three Chapters. Chapter 1 deals with the response of a simplified model of the arm, gimbal and gondola structure for the purpose of obtaining dynamic response factors to be associated with the arm. Chapter 2 deals briefly with a simplified model of the same system for the purpose of obtaining dynamic response factors to be associated with the gimbal ring and to justify the simplifications implicit in the model used in Chapter 1. In Chapter 3, the rigid body kinematic equations are studied in order to develop relations between the forcing functions utilized in Chapters 1 and 2 and the motion parameters of the kinematic analysis. Using these relations, the dynamic response factors tabulated in Chapters 1 and 2 in terms of the generalized forcing functions may be interpreted in terms of the motion parameters.

The following assumptions have been made in order to obtain a solution within a reasonable time period for the preliminary study:

    1. The gondola will be considered to be a rigid body;

2. The gimbal ring will also be considered to be a rigid body in the analysis of Chapter 1, and will be considered to be a simple spring-mass system in the analysis of Chapter 2;

3. The arm will be studied as another simple spring-mass system, acting as a uniform cantilever with a mass at its tip;

4. Small deflection, linear theory will be employed throughout the analysis;

5. Structural damping will be ignored for conservatism;

6. Lateral (tangential), vertical and torsional modes will be considered to be uncoupled and studied separately.

In defense of the apparent over-simplification of the complex system, which would be avoided if time permitted, it is to be observed that the simplifications are not as restrictive as they appear. The reason is that: 1) both the gondola and the gimbal ring are relatively rigid compared to the arm. Since the forcing functions have time durations that are not common in periodicity with the more rigid system but more common with the arm, simplification is permissible (as shown in Chapter 2 and Appendix 6); 2) The lumped parameter method is a conventional method of analysis; 3) Structural damping is small and the effect on transient motion is thus negligible; 4) An analysis (Appendix 3) has shown that coupling between modes does not occur; and 5) Neglecting the extensional mode is a conventional assumption and dynamic augmentation of centrifugal force is negligible.

The preliminary study of the structural dynamic response of the MSC centrifuge to transient loads resulting from gimbal-controlled motions has been completed. The results of the study are summarized in tables of dynamic response factors for lateral, torsional and vertical modes for three types of generalized impulsive loading functions: 1) a square pulse, 2) a saw-tooth ramp, and 3) a half sine pulse. In Chapter 3 of the study, the rigid body kinematic equations have been analyzed so that these generalized loading functions used in the analysis may be interpreted in terms of the motion parameters.

Contents

Abstract   iii

List of Tables   ix

List of Figures   x

Nomenclature   xi

1. Introduction   1

    1.1   Overview   1

    1.2   Summary   2

2. Arm Response   4

    2.1   Overview   4

    2.2   Summary- Equations of Motion   7

    2.3   Y Axis Analysis   7

    2.4   Z Axis Analysis   30

    2.5   X Axis Analysis   40

    2.6   Mass and Inertia Properties   46

    2.7   Numerical Calculations – X axis   47

    2.8   Numerical Calculations – Z axis   55

    2.9   Numerical Calculations – Y axis   62

3. Gimbal Response   72

3.1 Overview   72

3.2 Analysis   73

3.3 Numerical Calculations   79

4. Forcing Functions   82

    4.1 Overview   82

    4.2 Analysis   83

    4.3 Numerical Calculations   85

    4.4 Restraints on the System   87

    4.5 Analysis for Transient Loads   89

    4.6 Summary- Transient Loads Acting on the Arm   94

    4.7 Summary- Transient Loads Acting on the Gimbal Ring   95

    4.8 Study of Harmonic Forcing Functions   96

5. Conclusions   97

Bibliography

Appendix 1: Derivation of Y Axis Equations of Motion for Arm Model

Appendix 2: Derivation of Flexibility Matrix for the Arm

Appendix 3: Investigation of Coupling Between Vertical, Lateral
                 and Torsional Modes

Appendix 4: Computer Code and Results of Transient Response
                 Analysis

Appendix 5: Flexibility Coefficients of Gimbal Ring and Arm System

Appendix 6: Analysis of Coupling Effects of Arm and Gimbal Ring

Appendix 7: Photograph of the MSC Centrifuge

List of Tables

Table I	Mass and Inertia Properties of the Centrifuge	45
Table II	Times at Which Maximum Response Occur- Square Step	47
Table III	Dynamic Response Factors- Square Step	48
Table IV	Times at Which Maximum Response Occur- Ramp Input	50
Table V	Dynamic Response Factors- Square Step	51
Table VI	Times at Which Maximum Response Occur- Half Sine	52
Table VII	Dynamic Response Factors- Half Sine	53
Table VIII	Transient loads Acting on the Arm	93
Table IX	Transient loads Acting on the Gimbal Ring	94

List of Figures

Number	Description	Page Number
Figure 2.1.1	Schematic View of MSC Centrifuge	3
Figure 2.1.2	Idealized Mathematical Model	4
Figure 3.2.1	Idealized System	71
Figure 3.2.2	Mathematical Model	72
Figure A.1	Model in Y Plane	Appendix 1, pg. 2
Figure A.2	Nodal Notation	Appendix 2, pg. 2
Figure A.3	Case 2 Deflection Summary	Appendix 2, pg. 2
Figure A.4	Case 1 Deflection Summary	Appendix 2, pg. 3
Figure A.5	Case 3 Deflection Summary	Appendix 2, pg. 4
Figure A.6	Case 4 Deflection Summary	Appendix 2, pg. 5
Figure A.7	M_x Deflection Summary	Appendix 2, pg. 6
Figure A.8	M_y Deflection Summary	Appendix 2, pg. 7
Figure A.7	Photograph of MSC Centrifuge	Appendix 7, pg. 2

Nomenclature

Acronyms

MSC   Manned Spacecraft Center

Roman Symbols

A,B,C,D,E	Constants
F	Forcing Functions
I	Moment of Inertia
i and j	Subscripts
K	Spring Constant
L	Laplace Transform
M or m	Mass or Subscript for Mass
o	Subscript for Time Constant
r	Subscript Denoting Arm
s	Laplacian Operator
t	Time
T	Torque or Moment
X or x	Torsional Axis or Subscript for Longitudinal Axis

Y or y	Lateral Axis or Subscript for Lateral axis
Z or z	Vertical Axis or Subscript for Vertical Axis

Greek Symbols

α	Flexibility Coefficient
β	Integration Variable
γ	Dynamic Response Factor
δ	Displacement
Δ	Characteristic Determinant
Θ	Roll Angle
Ψ	Yaw Angle and Yaw Axis
Φ	Pitch Angle
ς	Eigenvalue
ω	Frequency
Θ_x	Angle about Torsional Axis
Θ_y	Angle about Roll Axis
Θ_z	Angle about Yaw Axis

Chapter 1 Introduction

1.1 Overview

This preliminary study is intended as an investigation into the structural dynamic behavior of the National Aeronautics and Space Administration (NASA) Manned Space Center (MSC) centrifuge used for training astronauts. In particular, the dynamic response to transient forcing functions that occur under imposed gondola rotations during astronaut training is predicted in the study.

The study is in three parts. Part A investigates the response of a simplified model of the arm, gimbal and gondola structure for the purpose of obtaining dynamic response factors to be associated with the arm. Part B deals briefly with a simplified model of the same system for the purpose of obtaining dynamic response factors to be associated with the gimbal ring and to justify the simplifications implicit in the model employed in Part A. In Part C, the rigid body kinematic equations are studied in order to obtain relations between the forcing functions used in Parts A and B and the motion parameters of the kinematic analysis. Using these relations, the dynamic response factors tabulated in Parts A and B in terms of the generalized forcing functions may be interpreted in terms of the motion parameters.

The following assumptions have been made in order to obtain a solution within the time available for the preliminary study:

1. The gondola will be considered to be a rigid body;

2. The gimbal ring will also be considered to be a rigid body in the analysis of Part A and it will be considered to be a simple spring-mass system in the analysis of Part B;

3. The arm will be considered to be a simple spring-mass system, acting as a uniform cantilever with a portion of its mass located at the tip of the arm;

4. Small deflection, linear theory will be employed throughout the analysis;

5. Structural damping will be ignored for conservatism;

1

6. Lateral (tangential), vertical and torsional modes will be considered to be uncoupled and will be studied separately.

In defense of the apparent over-simplification of the system, which would be avoided in a study with unlimited time, it is to be observed that the simplifications are not as restrictive as they appear. The following factors must be considered in this regard:

1. Both the gondola and the gimbal ring are relatively rigid compared to the arm.

2. Since the forcing functions have time durations, which are not in common with the periodicity with the more rigid systems, but rather are in common with the period of the arm, simplification of the system is permissible (see Part B and Appendix 6 for a proof).

3. The lumped parameter technique is a conventional method of analysis and has been shown to be fairly accurate for the prediction of the response to transient forces.

4. An analysis (see Appendix 3) has shown that no coupling exists between the various modes of the simplified model during free vibration. A similar conclusion holds for the forced response if the forcing functions are considered to be independent of the secondary motions. Although it is possible that self-induced vibrations may occur as a result of coupling between kinematic forces and structural vibrations, an investigation of this phenomenon is beyond the scope of this preliminary investigation.

5. Neglecting the extensional mode is a conventional assumption. In the structure under investigation the dynamic augmentation of maximum centrifugal forces is negligible.

1.2 Summary

A preliminary study of the structural dynamic response of the NASDA-Houston MSC centrifuge to transient loads induced by rotations of the gondola during rotation of the centrifuge arm has been completed.

The results of the study are summarized in the "Tables of Dynamic Response Factors" for the arm

for lateral, vertical and torsional modes for three types of typical impulsive, generalized forcing

functions. They are:

1. Square step (better described as a rectangular pulse).

2. Ramp.

3. Half Sine.

In part C of the study, the rigid body kinematic equations have been analyzed so that the

generalized loading functions employed in the analysis may be interpreted in terms of the motion

parameters of the kinematic analysis.

Chapter 2 Arm Response

2.1   Overview

In this Chapter of the study the equations of motion derived in Appendix 1 are summarized and then utilized to determine the fundamental frequencies of free vibration. Then, the response of the gimbal arm, resulting from enforced transient motions of the gondola, is determined.

Figure 2.1.1 is a depiction of the centrifuge structure and Figure 2.1.2 shows a corresponding view of the idealized model of the structure.

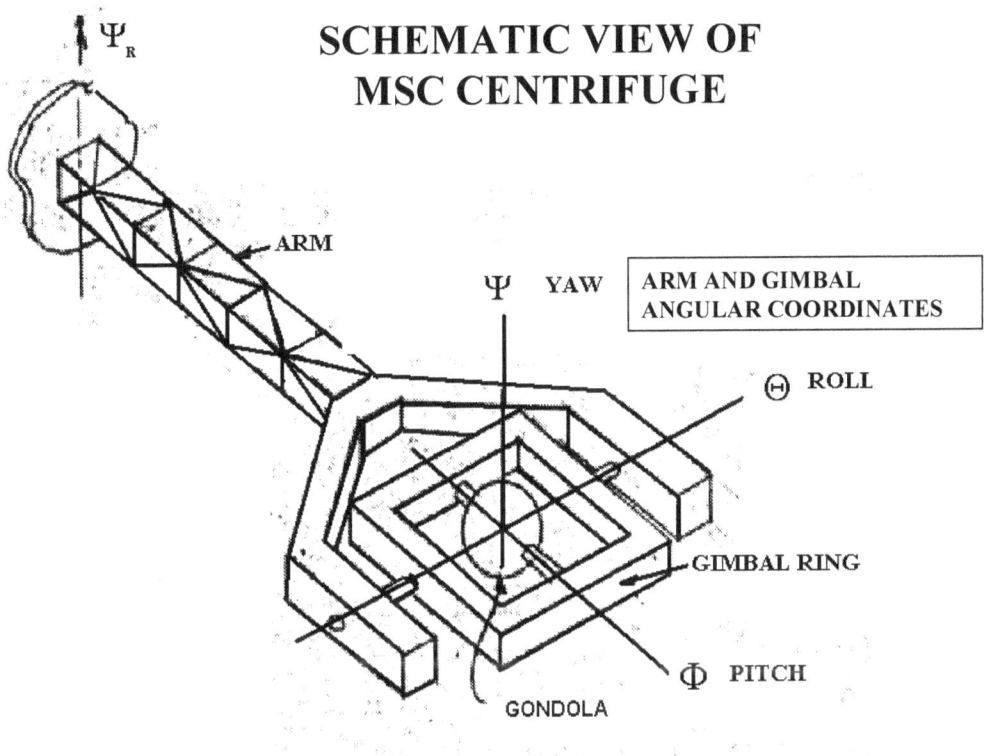

SCHEMATIC VIEW OF
MSC CENTRIFUGE

Ψ_R

ARM

Ψ YAW

ARM AND GIMBAL
ANGULAR COORDINATES

Θ ROLL

GIMBAL RING

Φ PITCH

GONDOLA

FIGURE 2.1.1

IDEALIZED MATHEMATICAL MODEL

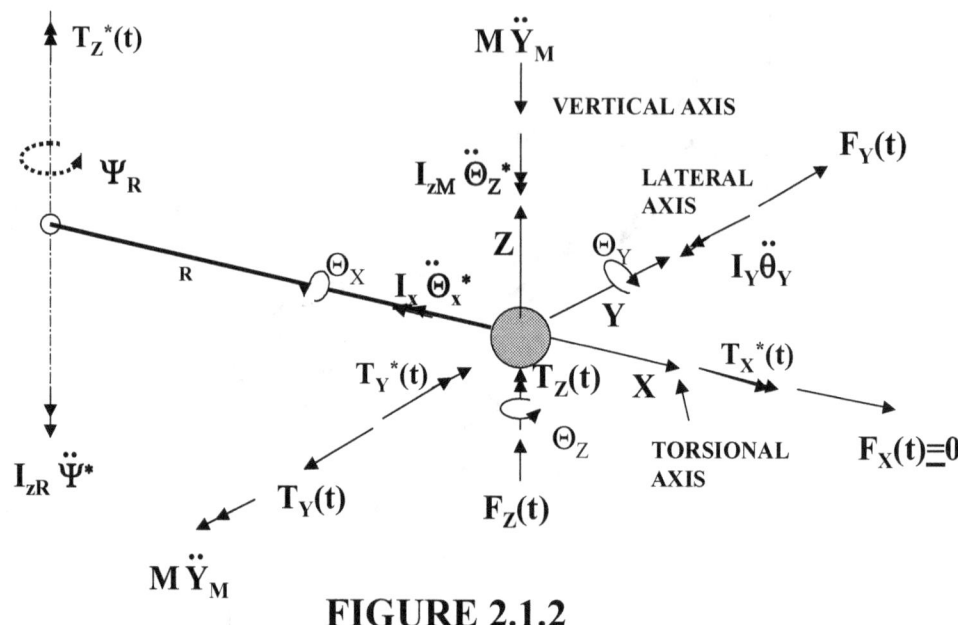

FIGURE 2.1.2

2.2   Summary – Equations of Motion

Y Axis:[1]

$$\begin{cases} MR\ddot{y}_M + I_{Z_M}\ddot{\theta}_Z + I_{Z_R}\ddot{\psi}^* &=& T_Z^*(t) + RF_y(t) + T_Z(t) \\ y_M - R\psi^* + \alpha_{11}M\ddot{y}_M + \alpha_{15}UI_{Z_M}\ddot{\theta}_Z &=& \alpha_{11}F_y(t) + \alpha_{15}T_Z(t) \\ \theta_Y + \psi^* + \alpha_{15}M\ddot{y}_M + \alpha_{55}I_{Z_M}\ddot{\theta}_Z &=& \alpha_{15}F_y(t) + \alpha_{55}T_Z(t) \end{cases} \quad (9)'$$

Equations for the two other axes may be written directly:

Z Axis:

$$\begin{cases} Z_M + \alpha_{22}M\ddot{Z}_M + \alpha_{24}I_{Y_M}\ddot{\theta}_Y &=& \alpha_{22}F_Z(t) + \alpha_{24}T_Y(t) \\ \theta_Y + \alpha_{42}M\ddot{Z}_M + \alpha_{44}I_{Y_M}\ddot{\theta}_Y &=& \alpha_{42}F_Z(t) + \alpha_{44}T_Y(t) \end{cases} \quad (10)$$

X Axis:

$$\left\{ \theta_X + \alpha_{33}I_{X_M}\ddot{\theta}_X \;=\; \alpha_{33}T_X(t) \right\} \quad (11)$$

Where the α_{ij} represent flexibility coefficients. (See Appendix 2)

2.3   Y –Axis Analysis

2.3.1 Solution for Y axis normal modes

Assuming a solution to the free vibration problem in the form:

$$\begin{cases} y_M = A_1\sin\omega t \\ \theta_Z = A_2\sin\omega t \\ \psi^* = A_3\sin\omega t \end{cases}$$

We obtain the following set of equations from (9):

$$-MR\omega^2 A_1 - I_{ZM}\omega^2 A_2 - I_{Z_R}\omega^2 A_3 \;=\; 0$$
$$(1 - \alpha_{11}M\omega^2)A_1 - \alpha_{15}I_{Z_M}\omega^2 A_2 - RA_3 \;=\; 0$$
$$-\alpha_{15}M\omega^2 A_1 + (1 - \alpha_{55}I_{Z_M}\omega^2)A_2 - A_3 \;=\; 0$$

---

[1] See Appendix 1 for derivation

Rearranging and using matrix notation:

$$
\begin{bmatrix}
\dfrac{1}{\omega^2} - a_{11} & a_{12} & \dfrac{1}{\omega^2} - a_{13} \\[2ex]
a_{21} & \dfrac{1}{\omega^2} - a_{22} & \dfrac{1}{\omega^2} - a_{23} \\[2ex]
a_{31} & a_{32} & a_{33}
\end{bmatrix}
\begin{bmatrix}
A_1 \\ A_2 \\ A_3
\end{bmatrix} = 0 \quad (12)
$$

Where,
$$
\left\{
\begin{array}{lll}
a_{11} = \alpha_{11}M; & a_{12} = -\alpha_{15}I_{Z_M}; & a_{13} = -R \\
a_{21} = \alpha_{15}M; & a_{22} = \alpha_{55}I_{Z_M}; & a_{23} = -1 \\
a_{31} = MR; & a_{32} = I_{Z_M}; & a_{33} = I_{Z_R}
\end{array}
\right\}
$$

The characteristic equation is given by the expansion of the determinant of the coefficient matrix set equal to zero. Thus,

$$
\left(\frac{1}{\omega^2} - a_{11}\right)\left(\frac{1}{\omega^2} - a_{22}\right)a_{33} + \frac{1}{\omega^2}\left(a_{13}a_{21}a_{32} + a_{12}a_{23}a_{31}\right) - \left(\frac{1}{\omega^2} - a_{22}\right)\left(\frac{1}{\omega^2}a_{13}\right)a_{31}
$$

$$
-\left(\frac{1}{\omega^2} - a_{11}\right)\left(\frac{1}{\omega^2}a_{23}\right)a_{32} - a_{12}a_{21}a_{33} = 0
$$

or,

$$
\left(\frac{1}{\omega^2}\right)^2\left[a_{33} - a_{13}a_{31} - a_{23}a_{32}\right] +
$$

$$
\left(\frac{1}{\omega^2}\right)\left[-a_{11}a_{33} - a_{22}a_{33} + a_{13}a_{21}a_{32} + a_{12}a_{23}a_{31} + a_{13}a_{22}a_{31} + a_{11}a_{23}a_{32}\right] - a_{12}a_{21}a_{33} + a_{11}a_{22}a_{33} = 0
$$

Denoting,
$$
\begin{aligned}
b_1 &= a_{33} - a_{13}a_{31} - a_{23}a_{32} \\
b_2 &= -a_{11}a_{33} - a_{22}a_{33} + a_{13}a_{22}a_{31} + a_{11}a_{23}a_{32} \\
b_3 &= -a_{11}a_{22}a_{33} - a_{12}a_{21}a_{33}
\end{aligned}
$$

Then: $\left(\dfrac{1}{\omega^2}\right) + A\left(\dfrac{1}{\omega^2}\right) + B = 0$ where, $\begin{aligned} A &= \dfrac{b_2}{b_1} \\ B &= \dfrac{b_3}{b_1} \end{aligned}$

Denoting $Z = \dfrac{1}{\omega^2}$

We find that $Z = -\dfrac{A}{2} \pm \sqrt{\left(\dfrac{A}{2}\right)^2 - B}$

Letting $Z_1 > Z_2$ where Z_1 and Z_2 are the characteristic roots, we have the natural frequencies: $\begin{cases} \omega_1 = Z_1^{-\frac{1}{2}} \\ \omega_2 = Z_2^{-\frac{1}{2}} \end{cases}$ (13)

The displacement ratios for each mode can be determined as follows:

Let $A_1 = 1$, for $\omega = \omega_i$

Then, from eqs. (12),

$$\frac{1}{\omega_i^2} - a_{11} + a_{12}A_2 + \frac{1}{\omega_i^2}a_{13}A_3 = 0$$

$$a_{21} + \left(\frac{1}{\omega_i^2} - a_{22}\right)A_2 + \frac{1}{\omega_i^2}a_{23}A_3 = 0$$

$$a_{31} + a_{32}A_2 + a_{33}A_3 = 0$$

From the last Eq. , $\qquad A_{2_i} = -\dfrac{a_{33}}{a_{32}}A_{3_i} - \dfrac{a_{31}}{a_{32}}$ (14)

Substitute into the second Eq. :

$$a_{21} + \left(\frac{1}{\omega_i^2} - a_{22}\right)\left(-\frac{a_{33}}{a_{32}}A_3 - \frac{a_{31}}{a_{32}}\right) + \frac{1}{\omega_i^2}a_{23}A_3 = 0$$

or, $a_{21} - \left(\dfrac{1}{\omega_i^2} - a_{22}\right)\dfrac{a_{31}}{a_{32}} = A_3\left[-\dfrac{a_{23}}{\omega_i^2} + \left(\dfrac{1}{\omega_i^2} - a_{22}\right)\dfrac{a_{33}}{a_{32}}\right] = 0$

or, $\qquad A_{3_i} = \dfrac{a_{21}a_{32} - (Z_i - a_{22})a_{31}}{(Z_i - a_{22})a_{33} - a_{23}a_{32}Z_i}$

$$A_{3_i} = \frac{a_{21}a_{32} + a_{22}a_{31} - a_{31}Z_i}{(a_{33} - a_{23}a_{32})Z_i - a_{22}a_{33}}$$ (15)

Of course, $\qquad A_{1_i} = 1$ (16)

Equations (14), (15), and (16) then define the displacement rations.

2.3.2 Solution For Y Axis Transient Response

Knowing the normal modes and frequencies of the system, the solution for the transient motions can be obtained in terms of these modes. Using a convenient numerical procedure (known as the phase-plane-delta method) where the forcing functions are complex time functions.

When the forcing functions are simple[2], a closed form solution using operation methods is possible.

Eqs. (9') are operated on by the Laplacian operator $s = \dfrac{d}{dt}$

Let:

$$\left[\begin{matrix} L\{y_M(t)\} = u_1(s) \\ L\{\theta_Z(t)\} = u_2(s) \\ L\{\psi^*(t)\} = u_3(s) \end{matrix}\right] \qquad \left[\begin{matrix} L\{T_Z^*(t)\} = f_1(s) \\ L\{F_y(t)\} = f_2(s) \\ L\{T_Z(t)\} = f_3(s) \end{matrix}\right]$$

Assume:

$$\left.\begin{matrix} \dot{y}_0 = \theta_{Z_0} = \psi_0 = 0 \\ \dot{y}_0 = R\dot{\psi}_0^* \\ \dot{\theta}_{Z_0} = \dot{\psi}_0^* \end{matrix}\right\}$$

Operating on Eqs. (9') : $\qquad L(\ddot{x}) = s^2 L(x) - sx_0 - \dot{x}_0$

Loading

$$\left.\begin{matrix} MR(s^2 u_1 - R\dot{\psi}_0^*) + I_{Z_M}(s^2 u_2 - \dot{\psi}_0^*) + I_{Z_R}(s^3 u_3 - \dot{\psi}_0^*) = & L[.....] \\ u_1 - Ru_3 + \alpha_{11}M(s^2 u_1 - R\dot{\psi}_0^*) + \alpha_{15}I_{Z_M}(s^2 u_2 - \dot{\psi}_0^*) = & L[.....] \\ u_2 - u_3 + \alpha_{15}M(s^2 u_1 - R\dot{\psi}_0^*) + \alpha_{55}I_{Z_M}(s^2 u_2 - \dot{\psi}_0^*) = & L[.....] \end{matrix}\right\} \qquad (17)$$

In matrix notation,

$$\begin{bmatrix} 1 + \alpha_{11}Ms^2 & -\alpha_{15}I_{Z_M}s^2 & -R \\ -\alpha_{15}Ms^2 & 1 + \alpha_{55}I_{Z_M}s^2 & -1 \\ MRs^2 & I_{Z_M}s^2 & I_{Z_M}s^2 \end{bmatrix} \begin{bmatrix} u_1 \\ u_2 \\ u_3 \end{bmatrix} = \begin{bmatrix} 1 \\ 0 \\ 0 \end{bmatrix} f_1(s) + \begin{bmatrix} R \\ \alpha_{11} \\ \alpha_{15} \end{bmatrix} f_2(s) + \begin{bmatrix} 1 \\ \alpha_{15} \\ \alpha_{55} \end{bmatrix} f_3(s)$$

$$+ \begin{bmatrix} MR^2 + I_{Z_M} + I_{Z_R} \\ \alpha_{11}MR + \alpha_{15}I_{Z_M} \\ \alpha_{15}MR + \alpha_{55}I_{Z_M} \end{bmatrix} \dot{\psi}_0^* = \begin{bmatrix} P_1 \\ P_2 \\ P_3 \end{bmatrix}$$

$$(18)$$

---

[2] Simple => Restricted to functions of class "A".

Using the notation of Eqs. (12), we have

$$\begin{bmatrix} 1+\alpha_{11}s^2 & -a_{12}s^2 & a_{13} \\ -a_{21}s^2 & 1+a_{22}s^2 & a_{23} \\ a_{31}s^2 & a_{32}s^2 & a_{33}s^2 \end{bmatrix} \begin{bmatrix} u_1 \\ u_2 \\ u_3 \end{bmatrix} = \begin{bmatrix} P_1 \\ P_2 \\ P_3 \end{bmatrix} \qquad (19)$$

Using Cramer's Rule:

$$u_1 = \frac{\begin{vmatrix} P_1 & -a_{12}s^2 & a_{13} \\ P_2 & 1+a_{22}s^2 & a_{23} \\ P_3 & a_{32}s^2 & a_{33}s^2 \end{vmatrix}}{\Delta}$$

$$u_1 = \frac{P_1\left[(1+a_{22}s^2)(a_{33}s^2)-a_{23}(a_{32}s^2)\right]}{\Delta} + \frac{P_2\left[a_{12}a_{33}s^4+a_{13}a_{32}s^2\right]}{\Delta}$$
$$+ \frac{P_3\left[-a_{12}a_{23}s^2-(1+a_{22}s^2)a_{13}\right]}{\Delta} \qquad (20)$$

$$u_2 = \frac{\begin{vmatrix} 1+a_{11}s^2 & P_1 & a_{13} \\ -a_{21}s^2 & P_2 & a_{23} \\ a_{31}s^2 & P_3 & a_{33}s^2 \end{vmatrix}}{\Delta}$$
$$= P_1\frac{\left[a_{21}a_{33}s^4+a_{23}a_{31}s^2\right]}{\Delta} + P_2\frac{\left[(1+a_{11}s^2)a_{33}s^2-a_{13}a_{31}s^2\right]}{\Delta}$$
$$+ P_3\frac{\left[-(1+a_{11}s^2)a_{23}-a_{13}a_{21}s^2\right]}{\Delta} \qquad (21)$$

$$u_3 = \frac{\begin{vmatrix} 1+a_{11}s^2 & -a_{12}s^2 & P_1 \\ -a_{21}s^2 & 1+a_{22}s^2 & P_2 \\ a_{31}s^2 & a_{32}s^2 & P_3 \end{vmatrix}}{\Delta}$$
$$= P_1\frac{\left[-a_{21}a_{32}s^4-(1+a_{22}s^2)a_{31}s^2\right]}{\Delta} + P_2\frac{\left[-(1+a_{11}s^2)a_{32}s^2-a_{12}a_{31}s^4\right]}{\Delta}$$
$$+ P_3\frac{\left[(1+a_{11}s^2)(1+a_{22}s^2)-a_{12}a_{21}s^4\right]}{\Delta} \qquad (22)$$

Where Δ = characteristic determinant = $p(s)$

11

$$p(s) = (1 + a_{11}s^2)(1 + a_{22}s^2)a_{33}s^2 - a_{12}s^2a_{23}a_{31}s^2 - a_{13}a_{32}s^2a_{21}s^2 - a_{13}a_{31}s^2(1 + a_{22}s^2)$$
$$- a_{23}a_{32}s^2(1 + a_{11}s^2) - a_{12}s^2a_{21}s^2a_{33}s^2$$
$$= s^6(a_{11}a_{22}a_{33} - a_{12}a_{21}a_{33}) + s^4(a_{22}a_{33} + a_{11}a_{33} - a_{12}a_{23}a_{31} - a_{13}a_{32}a_{21} - a_{13}a_{31}a_{22}$$
$$- a_{11}a_{23}a_{32}) + s^2(a_{33} - a_{13}a_{31} - a_{23}a_{32})$$

Letting:

$$b_3 = a_{11}a_{22}a_{33} - a_{12}a_{21}a_{33}$$
$$b_2 = -a_{11}a_{33} - a_{22}a_{33} + a_{13}a_{21}a_{32} + a_{12}a_{23}a_{31} + a_{13}a_{22}a_{31} + a_{11}a_{23}a_{32}$$
$$b_1 = a_{33} - a_{13}a_{31} - a_{23}a_{32}$$

Then:

$$p(s) = s^6(b_3) + s^4(-b_2) + s^2(b_1) = s^2\left[s^4(b_3) + s^2(-b_2) + (b_1)\right] \qquad (23)$$

Of course, the roots of $p(s)$ correspond to the roots given by (13).

Letting $\tau = s^2$, and setting $p(s) = 0$, we have

$$\tau(b_3\tau^2 - b_2\tau + b_1) = 0 \qquad\qquad \text{Denoting} \begin{cases} A' = -b_2 \Big/ b_3 \\ B' = b_1 \Big/ b_3 \end{cases}$$

We have

$$\tau^2 + A'\tau + B' = 0, \quad \tau = 0$$
with Eigenvalues:

$$\tau = -\frac{A'}{2} \pm \sqrt{\left(\frac{A'}{2}\right)^2 - B'} \quad ; \quad \tau = 0 \qquad (24)$$

Denote: $\begin{cases} \tau_1 < \tau_2 < 0 \\ s_i^2 = |\tau_i| \end{cases}$

Expanding $p(s)$ in terms of its roots:

$$p(s) = b_3 s^2 (s^2 + s_1^2)(s^2 + s_2^2) \qquad (25)$$

Thus, Eq. (20) can be expanded in partial fractions as follows:

$$b_3 u_1 = P_1 \left\{ \frac{A_{11}}{s^2} + \frac{A_{12}}{s^2 + s_1^2} + \frac{A_{13}}{s^2 + s_2^2} \right\}$$

$$+ P_2 \left\{ \frac{A_{21}}{s^2} + \frac{A_{22}}{s^2 + s_1^2} + \frac{A_{23}}{s^2 + s_2^2} \right\} \tag{26}$$

$$+ P_3 \left\{ \frac{A_{31}}{s^2} + \frac{A_{32}}{s^2 + s_1^2} + \frac{A_{33}}{s^2 + s_2^2} \right\}$$

Now,

$$A_{11}(s^2 + s_1^2)(s^2 + s_2^2) + A_{12}(s^2)(s^2 + s_2^2) + A_{13}(s^2)(s^2 + s_1^2) = a_{22}a_{33}s^4 + s^2(a_{33} - a_{23}a_{32})$$

$$A_{21}(s^2 + s_1^2)(s^2 + s_2^2) + A_{22}(s^2)(s^2 + s_2^2) + A_{23}(s^2)(s^2 + s_1^2) = a_{12}a_{33}s^4 + a_{13}a_{32}s^2 \tag{27}$$

$$A_{31}(s^2 + s_1^2)(s^2 + s_2^2) + A_{32}(s^2)(s^2 + s_2^2) + A_{33}(s^2)(s^2 + s_1^2) = -(a_{12}a_{23} + a_{13}a_{22}) - a_{13}$$

Using the values $s^2 = 0, \quad -s_1^2, \quad -s_2^2$ successively in (27), we obtain

$$P_1 \left\{ \begin{array}{lll} s^2 = 0: & A_{11}s_1^2 s_2^2 = 0, & A_{11} = 0 \\ s^2 = -s_1^2: & A_{12}(-s_1^2)(-s_1^2 + s_2^2) = a_{22}a_{33}s_1^4 - s_1^2(a_{33} - a_{23}a_{32}) \\ s^2 = -s_2^2: & A_{13}(-s_2^2)(-s_2^2 + s_1^2) = a_{22}a_{33}s_2^4 - s_2^2(a_{33} - a_{23}a_{32}) \end{array} \right\}$$

$$A_{12} = \frac{s_1^2[a_{22}a_{33}s_1^2 - (a_{33} - a_{23}a_{32})]}{-s_1^2(s_2^2 - s_1^2)} = -\frac{a_{22}a_{33}s_1^2 - (a_{33} - a_{23}a_{32})}{s_2^2 - s_1^2}$$

$$A_{13} = \frac{s_2^2[a_{22}a_{33}s_2^2 - (a_{33} - a_{23}a_{32})]}{-s_2^2(s_1^2 - s_2^2)} = +\frac{a_{22}a_{33}s_2^2 - (a_{33} - a_{23}a_{32})}{s_2^2 - s_1^2}$$

$$P_2 \left\{ \begin{array}{lll} s^2 = 0: & A_{21}s_1^2 s_2^2 = 0, & A_{21} = 0 \\ s^2 = -s_1^2: & A_{22}(-s_1^2)(-s_1^2 + s_2^2) = a_{12}a_{33}s_1^4 - s_1^2 a_{13}a_{32} \\ s^2 = -s_2^2: & A_{23}(-s_2^2)(-s_2^2 + s_1^2) = a_{12}a_{33}s_2^4 - s_2^2 a_{13}a_{32} \end{array} \right\}$$

$$A_{22} = \frac{s_1^2(a_{12}a_{33}s_1^2 - a_{13}a_{32})}{-s_1^2(s_2^2 - s_1^2)} = -\frac{a_{12}a_{33}s_1^2 - a_{13}a_{32}}{s_2^2 - s_1^2}$$

$$A_{23} = \frac{s_2^2(a_{12}a_{33}s_2^2 - a_{13}a_{32})}{-s_2^2(-s_2^2 + s_1^2)} = +\frac{a_{12}a_{33}s_2^2 - a_{13}a_{32}}{s_2^2 - s_1^2}$$

$$P_3 \left\{ \begin{array}{ll} s^2 = 0: & A_{31}s_1^2 s_2^2 = -a_{13} \\ s^2 = -s_1^2: & A_{32}(-s_1^2)(-s_1^2 + s_2^2) = (a_{12}a_{23} + a_{13}a_{22})s_1^2 - a_{13} \\ s^2 = -s_2^2: & A_{33}(-s_2^2)(-s_2^2 + s_1^2) = (a_{12}a_{23} + a_{13}a_{22})s_2^2 - a_{13} \end{array} \right\}$$

$$A_{31} = -\frac{a_{13}}{s_1^2 s_2^2}$$

$$A_{32} = \frac{s_1^2(a_{12}a_{23} + a_{13}a_{22}) - a_{13}}{-s_1^2(s_2^2 - s_1^2)} = -\frac{a_{12}a_{23} + a_{13}a_{22}}{s_2^2 - s_1^2} + \frac{a_{13}}{s_1^2(s_2^2 - s_1^2)}$$

$$A_{33} = \frac{s_2^2(a_{12}a_{23} + a_{13}a_{22}) - a_{13}}{-s_2^2(-s_2^2 - s_1^2)} = +\frac{a_{12}a_{23} + a_{13}a_{22}}{s_2^2 - s_1^2} - \frac{a_{13}}{s_2^2(s_2^2 - s_1^2)}$$

Note: The coefficients A_{ij} are constant!

Similarly, expanding Eq. (21) in partial fractions:

$$b_3 u_2 = P_1 \left\{ \frac{B_{11}}{s^2} + \frac{B_{12}}{s^2 + s_1^2} + \frac{B_{13}}{s^2 + s_2^2} \right\} + P_2 \left\{ \frac{B_{21}}{s^2} + \frac{B_{22}}{s^2 + s_1^2} + \frac{B_{23}}{s^2 + s_2^2} \right\}$$
$$+ P_3 \left\{ \frac{B_{31}}{s^2} + \frac{B_{32}}{s^2 + s_1^2} + \frac{B_{33}}{s^2 + s_2^2} \right\} \tag{28}$$

Now,

$$B_{11}(s^2 + s_1^2)(s^2 + s_2^2) + B_{12}(s^2)(s^2 + s_2^2) + B_{13}(s^2)(s^2 + s_1^2) = a_{21}a_{33}s^4 + a_{23}a_{31}s^2 \tag{29}$$
$$B_{21}(s^2 + s_1^2)(s^2 + s_2^2) + B_{22}(s^2)(s^2 + s_2^2) + B_{23}(s^2)(s^2 + s_1^2) = s^4(a_{11}a_{33}) + s^2(a_{33} - a_{13}a_{31})$$
$$B_{31}(s^2 + s_1^2)(s^2 + s_2^2) + B_{32}(s^2)(s^2 + s_2^2) + B_{33}(s^2)(s^2 + s_1^2) = -s^2(a_{11}a_{23} + a_{13}a_{21}) - a_{23}$$

Using the values $s^2 = 0, \quad s^2 = -s_1^2, \quad s^2 = -s_2^2$ successively in Eqs. (29), we have

$$P_1 \begin{cases} s^2 = 0: & B_{11}s_1^2 s_2^2 = 0, & B_{11} = 0 \\ s^2 = -s_1^2: & B_{12}(-s_1^2)(-s_1^2 + s_2^2) = a_{21}a_{33}s_1^4 - a_{23}a_{31}s_1^2 \\ s^2 = -s_2^2: & B_{13}(-s_2^2)(-s_2^2 + s_1^2) = a_{21}a_{33}s_2^4 - a_{23}a_{31}s_2^2 \end{cases}$$

$$B_{12} = \frac{s_1^2[a_{21}a_{33}s_1^2 - a_{23}a_{31}]}{-s_1^2(s_2^2 - s_1^2)} = -\frac{a_{21}a_{33}s_1^2 - a_{23}a_{31}}{s_2^2 - s_1^2}$$

$$B_{13} = \frac{s_2^2[a_{21}a_{33}s_2^2 - a_{23}a_{31}]}{s_2^2(s_2^2 - s_1^2)} = \frac{a_{21}a_{33}s_2^2 - a_{23}a_{31}}{s_2^2 - s_1^2}$$

$$P_2 \begin{cases} s^2 = 0: & B_{21}s_1^2 s_2^2 = 0, & B_{21} = 0 \\ s^2 = -s_1^2: & B_{12}(-s_1^2)(-s_1^2 + s_2^2) = a_{11}a_{33}s_1^4 - s_1^2(a_{33} - a_{13}a_{31}) \\ s^2 = -s_2^2: & B_{23}(-s_2^2)(-s_2^2 + s_1^2) = a_{11}a_{33}s_2^4 - s_2^2(a_{33} - a_{13}a_{31}) \end{cases}$$

$$B_{22} = \frac{s_1^2[a_{11}a_{33}s_1^2 - (a_{33} - a_{13}a_{31})]}{-s_1^2(s_2^2 - s_1^2)} = -\frac{a_{11}a_{33}s_1^2 - (a_{33} - a_{13}a_{31})}{s_2^2 - s_1^2}$$

14

$$B_{23} = \frac{s_2^2[a_{11}a_{33}s_2^2 - (a_{33} - a_{13}a_{31})]}{s_2^2(s_2^2 - s_1^2)} = \frac{a_{11}a_{33}s_2^2 - (a_{33} - a_{13}a_{31})}{s_2^2 - s_1^2}$$

$$P_3 \begin{cases} s^2 = 0: & B_{21}s_1^2 s_2^2 = a_{23}, \\ s^2 = -s_1^2: & B_{32}(-s_1^2)(-s_1^2 + s_2^2) = s_1^2(a_{11}a_{23} + a_{13}a_{21}) - a_{23} \\ s^2 = -s_2^2: & B_{33}(-s_2^2)(-s_2^2 + s_1^2) = s_2^2(a_{11}a_{23} - a_{13}a_{21}) - a_{23} \end{cases}$$

$$B_{31} = -\frac{a_{23}}{s_1^2 s_2^2}$$

$$B_{32} = \frac{s_1^2(a_{11}a_{23} + a_{13}a_{21}) - a_{23}}{-s_1^2(s_2^2 - s_1^2)} = -\frac{a_{11}a_{23} + a_{13}a_{32}}{s_2^2 - s_1^2} + \frac{a_{23}}{s_1^2(s_2^2 - s_1^2)}$$

$$B_{33} = \frac{s_2^2(a_{11}a_{23} + a_{13}a_{21}) - a_{23}}{s_2^2(s_2^2 - s_1^2)} = \frac{a_{11}a_{23} + a_{13}a_{32}}{s_2^2 - s_1^2} - \frac{a_{23}}{s_2^2(s_2^2 - s_1^2)}$$

Again, expanding Eq. (22) in partial fractions:

$$b_3 u_3 = P_1 \left\{ \frac{C_{11}}{s^2} + \frac{C_{12}}{s^2 + s_1^2} + \frac{C_{13}}{s^2 + s_2^2} \right\} + P_2 \left\{ \frac{C_{21}}{s^2} + \frac{C_{22}}{s^2 + s_1^2} + \frac{C_{23}}{s^2 + s_2^2} \right\}$$

$$+ P_3 \left\{ \frac{C_{31}}{s^2} + \frac{C_{32}}{s^2 + s_1^2} + \frac{C_{33}}{s^2 + s_2^2} \right\} \tag{30}$$

Now,

$$C_{11}(s^2 + s_1^2)(s^2 + s_2^2) + C_{12}(s^2)(s^2 + s_2^2) + C_{13}(s^2)(s^2 + s_1^2) = -s^4(a_{21}a_{33} + a_{22}a_{31}) - a_{31}s^2$$

$$C_{21}(s^2 + s_1^2)(s^2 + s_2^2) + C_{22}(s^2)(s^2 + s_2^2) + C_{23}(s^2)(s^2 + s_1^2) = -s^4(a_{11}a_{32} + a_{12}a_{31}) - a_{32}s^2$$

$$C_{31}(s^2 + s_1^2)(s^2 + s_2^2) + C_{32}(s^2)(s^2 + s_2^2) + C_{33}(s^2)(s^2 + s_1^2) =$$

$$s^4(a_{11}a_{22} - a_{12}a_{21}) + s^2(a_{11} + a_{22}) + 1 \tag{31}$$

Using the values $s^2 = 0$, $s^2 = -s_1^2$, $s^2 = -s_2^2$ successively in Eqs. (31), we have

$$P_1 \begin{cases} s^2 = 0: & C_{11}s_1^2 s_2^2 = 0, \qquad C_{11} = 0 \\ s^2 = -s_1^2: & C_{12}(-s_1^2)(-s_1^2 + s_2^2) = -s_1^4(a_{21}a_{32} + a_{22}a_{31}) + a_{31}s_1^2 \\ s^2 = -s_2^2: & C_{13}(-s_2^2)(-s_2^2 + s_1^2) = -s_2^4(a_{21}a_{32} - a_{22}a_{31}) + a_{31}s_2^2 \end{cases}$$

$$C_{12} = \frac{-s_1^2[s_1^2(a_{21}a_{32} - a_{22}a_{31}) - a_{31}]}{-s_1^2(s_2^2 - s_1^2)} = \frac{s_1^2(a_{21}a_{32} + a_{22}a_{31}) - a_{31}}{s_2^2 - s_1^2}$$

$$C_{13} = \frac{-s_2^2[s_2^2(a_{21}a_{32} - a_{22}a_{31}) - a_{31}]}{s_2^2(s_2^2 - s_1^2)} = -\frac{s_2^2(a_{21}a_{32} + a_{22}a_{31}) - a_{31}}{s_2^2 - s_1^2}$$

$$P_2 \begin{cases} s^2 = 0: & C_{21}s_1^2 s_2^2 = 0, \qquad C_{21} = 0 \\ s^2 = -s_1^2: & C_{22}(-s_1^2)(-s_1^2 + s_2^2) = s_1^4(a_{11}a_{32} + a_{12}a_{31}) + a_{32}s_1^2 \\ s^2 = -s_2^2: & C_{23}(-s_2^2)(-s_2^2 + s_1^2) = s_1^4(a_{11}a_{32} + a_{12}a_{31}) + a_{32}s_2^2 \end{cases}$$

15

$$C_{22} = \frac{-s_1^2[s_1^2(a_{11}a_{32} - a_{12}a_{31}) - a_{32}]}{-s_1^2(s_2^2 - s_1^2)} = \frac{s_1^2(a_{11}a_{32} + a_{12}a_{31}) - a_{32}}{s_2^2 - s_1^2}$$

$$C_{23} = \frac{-s_2^2[s_2^2(a_{11}a_{32} - a_{12}a_{31}) - a_{32}]}{s_2^2(s_2^2 - s_1^2)} = -\frac{s_2^2(a_{11}a_{32} + a_{12}a_{31}) - a_{32}}{s_2^2 - s_1^2}$$

$$P_3 \begin{cases} s^2 = 0: & C_{31}s_1^2s_2^2 = 1; & C_{31} = \dfrac{1}{s_1^2 s_2^2} \\ s^2 = -s_1^2: & C_{32}(-s_1^2)(-s_1^2 + s_2^2) = s_1^4(a_{11}a_{22} - a_{12}a_{21}) - s_1^2(a_{11} + a_{22}) + 1 \\ s^2 = -s_2^2: & C_{33}(-s_2^2)(-s_2^2 + s_1^2) = s_2^4(a_{11}a_{22} - a_{12}a_{21}) - s_2^2(a_{11} + a_{22}) + 1 \end{cases}$$

$$C_{32} = \frac{s_1^4(a_{11}a_{22} - a_{12}a_{21}) - s_1^2(a_{11} + a_{22}) + 1}{-s_1^2(s_2^2 - s_1^2)} = \frac{-s_1^2(a_{11}a_{22} - a_{12}a_{21})}{s_2^2 - s_1^2} + \frac{a_{11} + a_{22}}{s_2^2 - s_1^2} - \frac{1}{s_1^2(s_2^2 - s_1^2)}$$

$$C_{33} = \frac{s_2^4(a_{11}a_{22} - a_{12}a_{21}) - s_2^2(a_{11} + a_{22}) + 1}{s_2^2(s_2^2 - s_1^2)} = \frac{s_2^2(a_{11}a_{22} - a_{12}a_{21})}{s_2^2 - s_1^2} - \frac{a_{11} + a_{22}}{s_2^2 - s_1^2} + \frac{1}{s_2^2(s_2^2 - s_1^2)}$$

The constants A_{ij}, B_{ij}, C_{ij} are now defined.

Recall from Eq. (18) that

$$\begin{aligned} P_1 &= f_1(s) + Rf_2(s) + f_3(s) + K_1\dot{\psi}_0^* \\ P_2 &= \alpha_{11}f_2(s) + \alpha_{15}f_3(s) + K_2\dot{\psi}_0^* \\ P_3 &= \alpha_{15}f_2(s) + \alpha_{55}f_3(s) + K_3\dot{\psi}_0^* \end{aligned} \tag{32}$$

Where
$$\begin{aligned} K_1 &= MR^2 + I_{Z_M} + I_{Z_R} \\ K_2 &= \alpha_{11}MR + \alpha_{15}I_{Z_M} \\ K_3 &= \alpha_{15}MR + \alpha_{55}I_{Z_M} \end{aligned} \tag{33}$$

The terms in Eqs. (32) involving K_i yield the system response due to a step velocity input $\dot{\psi}^*$, since the arm is assumed undeflected at t=0 (see pg. 6). A steady state $\dot{\psi}$ motion does not cause oscillations of the system; i.e., transient inputs will eventually damp out. (as seen from eqs. (6), pg. 5, terms involving the steady state velocity $\dot{\psi}$ drop out during the linearization of the motion equations). Since the K_i terms do not involve the transform variable "s", the inverse transforms can be evaluated immediately:

$$\left\{\begin{array}{l} L^{-1}\left\{\dfrac{k_{i1}K_i\dot{\psi}_0^*}{s^2}\right\} = k_{i1}K_i\dot{\psi}_0^* t \\[3mm] L^{-1}\left\{\dfrac{k_{i2}K_i\dot{\psi}_0^*}{s^2+s_1^2}\right\} = \dfrac{k_{i2}K_i\dot{\psi}_0^*}{s_1}\sin s_1 t \\[3mm] L^{-1}\left\{\dfrac{k_{i3}K_i\dot{\psi}_0^*}{s^2+s_2^2}\right\} = \dfrac{k_{i3}K_i\dot{\psi}_0^*}{s_2}\sin s_2 t \end{array}\right\} \tag{34}$$

For $\left\{\begin{array}{l} k = A,\,B,\,C \\ i = 1,\,2,\,3 \end{array}\right\}$

The inverse transforms of the other terms must be expressed in terms of the convolution theorem.

Recall that: $L^{-1}\{f(s)g(s)\} = \int_0^t F(\beta)G(t-\beta)d\beta \equiv \int_0^t F(t-\beta)G(\beta)d\beta$

Where $\qquad L^{-1}\{f(s)\} = F(t),\quad L^{-1}\{g(s)\} = G(t)$ (and F(t), G(t) are of class A)

Now, $L^{-1}\left\{\dfrac{ef_i(s)}{s^2}\right\} = \int_0^t eF_i(\beta)(t-\beta)d\beta$; $\qquad i = 1,\,2,\,3$

$$L^{-1}\left\{\dfrac{ef_i(s)}{s^2+s_N^2}\right\} = \int_0^t \dfrac{e}{s_N}F_i(\beta)\sin s_N(t-\beta)d\beta \qquad \left\{\begin{array}{l} i = 1,\,2,\,3 \\ N = 1,\,2 \end{array}\right\} \tag{35}$$

Where e = arbitrary constants (Eqs. 18)

The forcing function $F_i(t)$ are, from pg. 6,

$$\begin{array}{l} F_1(t) = L^{-1}\{f_1(s)\} = T_z^*(t) \\[1mm] F_2(t) = L^{-1}\{f_2(s)\} = F_y(t) \\[1mm] F_3(t) = L^{-1}\{f_3(s)\} = T_Z(t) \end{array} \tag{36}$$

The type of functions that we will investigate will appear as follows:[3]

(a) Square steps:

---

[3] Note: Harmonic forcing functions are non-critical (see Part C)

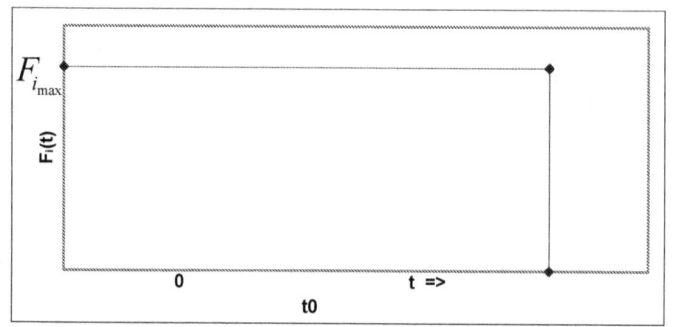

$$F_i(t) = F_{i_{max}}[1 - \alpha(t - t_0)] \qquad \text{where} \qquad \begin{array}{l} \alpha(t) = 0 \quad t < 0 \\ \alpha(t) = 1 \quad t \geq 0 \end{array}$$

(b) Ramps:

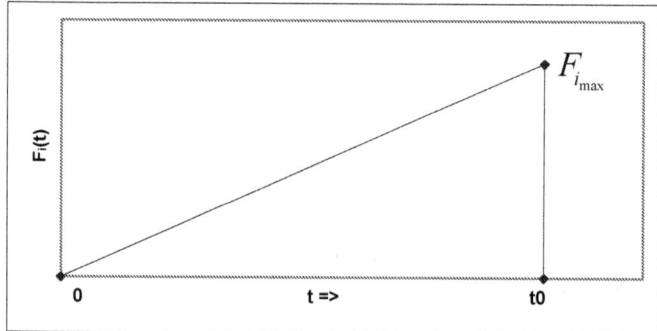

$$F_i(t) = \frac{F_{i_{max}}}{t_0} t[1 - \alpha(t - t_0)]$$

(b́) Negative Ramps:

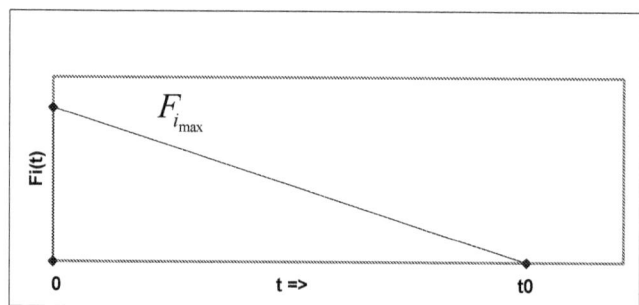

$$F_i(t) = F_{i_{max}}(1 - \frac{t}{t_0})[1 - \alpha(t - t_0)]$$

Note: this case is obtained from

solutions (a) and (b)

18

(c) Half sine waves:

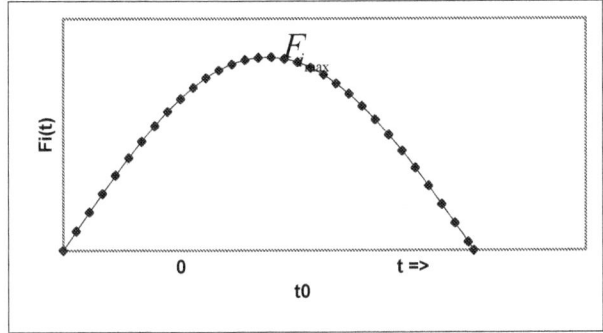

$$F_i(t) = F_{i_{max}} \sin\left(\frac{\pi t}{t_0}\right)\left[1 - \alpha(t - t_0)\right]$$

Eqs. (35) will now be integrated for each of the loading cases:

$$L^{-1}\left\{\frac{ef_i(s)}{s^2}\right\} = e\int_0^t F_i^\Diamond(\beta)(t - \beta)d\beta = e\int_0^t F_{i_{Max}}\left[1 - \alpha(\beta - t_0)\right](t - \beta)d\beta$$

$$= eF_{i_{Max}}\left\{\int_0^t(t - \beta)d\beta - \alpha(t - t_0)\int_t^{t_0}(t - \beta)d\beta\right\}$$

(a) $$= eF_{i_{Max}}\left\{\frac{t^2}{2} - \alpha(t - t_0)\frac{1}{2}(t - t_0)^2\right\} \qquad\qquad (37)$$

$$= eF_{i_{Max}}\left\{\begin{array}{ll}\dfrac{t^2}{2}, & t < t_0 \\[3ex] tt_0 - \dfrac{t_0^2}{2}, & t \geq t_0\end{array}\right\}$$

$$L^{-1}\left\{\frac{ef_i(s)}{s^2 + s_N^2}\right\} = \frac{e}{s_N}\int_0^t F_i^\Diamond(\beta)\sin s_N(t - \beta)d\beta = \frac{e}{s_N}\int_0^t F_{i_{Max}}\left[1 - \alpha(\beta - t_0)\right]\sin s_N(t - \beta)d\beta$$

$$= \frac{e}{s_N}F_{i_{Max}}\left\{\int_0^t \sin s_N(t - \beta)d\beta - \alpha(t - t_0)\int_{t_0}^t \sin s_N(t - \beta)d\beta\right\}$$

Let $\begin{array}{l}\omega = s_N(t - \beta) \\ d\omega = -s_N d\beta\end{array}$ when $\begin{array}{ll}\beta = t, & \omega = 0 \\ \beta = 0, & \omega = s_N t\end{array}$ $\begin{array}{ll}\beta = t_0, & \omega = s_N(t - t_0) \\ \beta = t, & \omega = 0\end{array}$

$$\int_{t_0}^t \sin s_N(t - \beta)d\beta = -\frac{1}{s_N}\int_{s_N t}^0 \sin\omega\, d\omega = \frac{1}{s_N}\int_0^{s_N t}\sin\omega\, d\omega = \frac{1}{s_N}[-\cos\ \omega]\Big|_0^{s_N t}$$

$$= \frac{1}{s_N}[1 - \cos s_N t]; \quad \int_{t_0}^t (\ )d\beta = \frac{1}{s_N}[-\cos\ \omega]\Big|_0^{s_N(t - t_0)} = \frac{1}{s_N}[1 - \cos s_N(t - t_0)]$$

19

$$\therefore L^{-1}\{\quad\} = \frac{eF_{i_{Max}}}{s_N}\left\{\begin{array}{ll}\dfrac{1}{s_N}[1-\cos s_N t], & t < t_0 \\[2mm] \dfrac{1}{s_N}[\cos s_N(t-t_0)-\cos s_N t], & t \geq t_0\end{array}\right\} \tag{38}$$

(b) $L^{-1}\left\{\dfrac{efF_i(s)}{s^2}\right\} = e\displaystyle\int_0^t F_i(\beta)(t-\beta)d\beta = \dfrac{eF_{i_{Max}}}{t_0}\int_0^t \beta[1-\alpha(\beta-t_0)](t-\beta)d\beta$

$$= \frac{eF_{i_{Max}}}{t_0}\left\{\int_0^t \beta(t-\beta)d\beta - \alpha(t-t_0)\int_{t_0}^t \beta(t-\beta)\right\}$$

$$= \frac{eF_{i_{Max}}}{t_0}\left\{\frac{t}{2}\left(\beta^2\Big|_0^t\right)-\frac{1}{3}\left(\beta^3\Big|_0^t\right)-\alpha(t-t_0)\left[\frac{t}{2}\left(\beta^2\Big|_0^t\right)-\frac{1}{3}\left(\beta^3\Big|_0^t\right)\right]\right\}$$

$$= \frac{eF_{i_{Max}}}{t_0}\left\{\left(\frac{t^3}{2}-\frac{t^3}{3}\right)-\alpha(t-t_0)\left[\frac{t}{2}(t^2-t_0^2)-\frac{1}{3}(t^3-t_0^3)\right]\right\}$$

$$= \frac{eF_{i_{Max}}}{t_0}\left\{\begin{array}{ll}\dfrac{1}{6}t^3, & t < t_0 \\[2mm] \dfrac{tt_0^2}{2}-\dfrac{t_0^3}{3}, & t \geq t_0\end{array}\right\} \tag{39}$$

$$L^{-1}\left\{\frac{ef_i(s)}{s^2+s_N^2}\right\} = \frac{e}{s_N}\int_0^t F_i^{\Delta}(\beta)\sin s_N(t-\beta)d\beta = \frac{eF_{i_{Max}}}{s_N t_0}\int_0^t \beta[1-\alpha(\beta-t_0)]\sin s_N(t-\beta)d\beta$$

$$= \frac{eF_{i_{Max}}}{s_N t_0}\left\{\int_0^t \beta\sin s_N(t-\beta)d\beta - \alpha(t-t_0)\int_{t_0}^t \beta\sin s_N(t-\beta)d\beta\right\}$$

Let $\begin{array}{l}\omega = s_N(t-\beta) \\ d\omega = -s_N d\beta\end{array}$ when $\begin{array}{l}\beta=t,\quad \omega=0,\quad \beta=\dfrac{s_N t-\omega}{s_N} \\[2mm] \beta=0,\quad \omega=s_N(t-t_0)\end{array}$

$$\int_{t_0}^t \beta\sin s_N(t-\beta)d\beta = -\int_{s_N(t-t_0)}^0 \left(\frac{s_N t-\omega}{s_N^2}\right)\sin\omega\, d\omega = \frac{1}{s_N^2}\int_0^{s_N(t-t_0)}(s_N t-\omega)\sin\omega\, d\omega$$

$$= \frac{1}{s_N^2}\left\{s_N t\left[-\cos\omega\Big|_0^{s_N(t-t_0)}\right]-\left[\sin\omega-\omega\cos\omega\Big|_0^{s_N(t-t_0)}\right]\right\}$$

$$= \frac{1}{s_N^2}\left\{s_N t[1-\cos s_N(t-t_0)]-[\sin s_N(t-t_0)-s_N(t-t_0)\cos s_N(t-t_0)]\right\}$$

$$= \frac{1}{s_N^2}\left\{s_N t[1-\cos s_N(t-t_0)]-\sin s_N(t-t_0)+s_N(t-t_0)\cos s_N(t-t_0)\right\}$$

$$\therefore L^{-1}\{\quad\} = \frac{eF_{i_{Max}}}{s_N^3 t_0}\left\{\begin{array}{l}s_N t(1-\cos s_N t)-\sin s_N t+s_N t\cos s_N t-\alpha(t-t_0)* \\ (s_N t[1-\cos s_N(t-t_0)]-\sin s_N(t-t_0)+s_N(t-t_0)\cos s_N(t-t_0))\end{array}\right\}$$

$$= \frac{eF_{i_{Max}}}{s_N^3 t_0}\left\{\begin{array}{ll}s_N t-\sin s_N t, & t < t_0 \\[2mm] s_N t\cos s_N(t-t_0)-\sin s_N t-s_N(t-t_0)\cos s_N(t-t_0)+\sin s_N(t-t_0), & t \geq t_0\end{array}\right\} \tag{40}$$

(c) $L^{-1}\left\{\dfrac{ef_i(s)}{s^2}\right\} = e\displaystyle\int_0^t F_i^{\Delta}(\beta)(t-\beta)d\beta = eF_{i_{Max}}\displaystyle\int_0^t \sin\left(\dfrac{\pi\beta}{t_0}\right)\left[1-\alpha(\beta-t_0)\right](t-\beta)d\beta$

$= eF_{i_{Max}}\left\{\displaystyle\int_0^t \sin\left(\dfrac{\pi\beta}{t_0}\right)(t-\beta)d\beta - \alpha(\beta-t_0)\displaystyle\int_{t_0}^t \sin\left(\dfrac{\pi\beta}{t_0}\right)(t-\beta)d\beta\right\}$

Let $\quad\begin{aligned}\omega &= \dfrac{\pi\beta}{t_0}\\[2mm] d\omega &= \dfrac{\pi\beta}{t_0}d\beta\end{aligned}\qquad$ when $\qquad\begin{aligned}\beta &= t, \quad \omega = \pi t/t_0, \qquad \beta = \dfrac{t_0\omega}{\pi}\\[2mm] \beta &= t_0, \quad \omega = \pi\end{aligned}$

$\displaystyle\int_{t_0}^t \sin\left(\dfrac{\pi\beta}{t_0}\right)(t-\beta)d\beta = \dfrac{t_0}{\pi}\displaystyle\int_{\pi}^{\pi/t_0} \sin\omega\left(t-\dfrac{t_0\omega}{\pi}\right)d\omega = \dfrac{t_0 t}{\pi}\displaystyle\int_{\pi}^{\pi/t_0} \sin\omega\, d\omega - \dfrac{t_0^2}{\pi}\displaystyle\int_{\pi}^{\pi}\omega\sin\omega\, d\omega$

$= \dfrac{t_0 t}{\pi}\left[-\cos\omega\right|_{\pi}^{\pi/t_0} - \dfrac{t_0^2}{\pi^2}\left[\sin\omega - \omega\cos\omega\right|_{\pi}^{\pi/t_0}$

$= \dfrac{t_0 t}{\pi}\left[-\cos \pi t/t_0 - 1\right] - \dfrac{t_0^2}{\pi^2}\left[\sin\dfrac{\pi t}{t_0} - \dfrac{\pi t}{t_0}\cos\dfrac{\pi t}{t_0} - \pi\right]$

$\omega = 0, \beta = 0$

$\displaystyle\int_0^t \sin\left(\dfrac{\pi\beta}{t_0}\right)(t-\beta)d\beta = \dfrac{t t_0}{\pi}\left[-\cos\omega\right|_0^{\pi/t_0} - \dfrac{t_0^2}{\pi^2}\left[\sin\omega - \omega\cos\omega\right|_0^{\pi/t_0}$

$= \dfrac{t t_0}{\pi}\left[1 - \cos\dfrac{\pi t}{t_0}\right] - \dfrac{t_0^2}{\pi^2}\left[\sin\dfrac{\pi t}{t_0} - \dfrac{\pi t}{t_0}\cos\dfrac{\pi t}{t_0}\right]$

$\therefore L^{-1}\{\quad\} = eF_{i_{Max}}\left\{\begin{aligned}&\dfrac{t_0 t}{\pi}\left[1-\cos\dfrac{\pi t}{t_0}\right] - \dfrac{t_0^2}{\pi^2}\left[\sin\dfrac{\pi t}{t_0} + \dfrac{\pi t}{t_0}\cos\dfrac{\pi t}{t_0}\right], \quad t < t_0\\[3mm] &\dfrac{2t_0 t}{\pi} - \dfrac{t_0^2}{\pi}, \qquad\qquad t \geq t_0\end{aligned}\right\}$ (41)

$L^{-1}\left\{\dfrac{ef_i(s)}{s^2 + s_N^2}\right\} = \dfrac{e}{s_N}\displaystyle\int_0^t F_i^{\Delta}(\beta)\sin s_N(t-\beta)\, d\beta$

$= \dfrac{e}{s_N}F_{i_{Max}}\displaystyle\int_0^t \sin\left(\dfrac{\pi\beta}{t_0}\right)\left[1-\alpha(\beta-t_0)\right]\sin s_N(t-\beta)\, d\beta$

$= \dfrac{eF_{i_{Max}}}{s_N}\left\{\displaystyle\int_0^t \sin\dfrac{\pi\beta}{t_0}\left[\sin s_N t\cos s_N\beta - \cos s_N t\sin s_N\beta\right]d\beta - \alpha(t-t_0)\displaystyle\int_{t_0}^t [\quad]\, d\beta\right\}$

$$
= \frac{eF_{i_{Max}}}{s_N} \left\{
\begin{array}{l}
\sin s_N t \int_0^t \sin \frac{\pi\beta}{t_0} \cos s_N \beta \, d\beta - \cos s_N t \int_0^t \sin \frac{\pi\beta}{t_0} \sin s_N \beta \, d\beta \\[2ex]
- \alpha(t - t_0) \left[\sin s_N t \int_{t_0}^t \sin \frac{\pi\beta}{t_0} \cos s_N \beta \, d\beta - \cos s_N t \int_{t_0}^t \sin \frac{\pi\beta}{t_0} \sin s_N \beta \, d\beta \right]
\end{array}
\right\}
$$

$$
= -\frac{eF_{i_{Max}}}{s_N} \left\{
\begin{array}{l}
+ \frac{\sin s_N t}{2} \left[\frac{\cos\left(\frac{\pi}{t_0} - s_N\right)\beta}{\frac{\pi}{t_0} - s_N} + \frac{\cos\left(\frac{\pi}{t_0} + s_N\right)\beta}{\frac{\pi}{t_0} + s_N} \right]_0^t + \frac{\cos s_N t}{2} \left[\frac{\sin\left(\frac{\pi}{t_0} - s_N\right)\beta}{\frac{\pi}{t_0} - s_N} - \frac{\sin\left(\frac{\pi}{t_0} + s_N\right)\beta}{\frac{\pi}{t_0} + s_N} \right]_0^t \\[3ex]
- \alpha(t - t_0) \frac{\sin s_N t}{2} \left[\frac{\cos\left(\frac{\pi}{t_0} - s_N\right)\beta}{\frac{\pi}{t_0} - s_N} + \frac{\cos\left(\frac{\pi}{t_0} + s_N\right)\beta}{\frac{\pi}{t_0} + s_N} \right]_{t_0}^t + \frac{\cos s_N t}{2} \left[\frac{\sin\left(\frac{\pi}{t_0} - s_N\right)\beta}{\frac{\pi}{t_0} - s_N} - \frac{\sin\left(\frac{\pi}{t_0} + s_N\right)\beta}{\frac{\pi}{t_0} + s_N} \right]_{t_0}^t
\end{array}
\right\}
$$

$$
= -\frac{eF_{i_{Max}}}{2s_N} \left\{
\begin{array}{l}
\sin s_N t \left[\frac{\cos\left(\frac{\pi}{t_0} - s_N\right)t}{\frac{\pi}{t_0} - s_N} + \frac{\cos\left(\frac{\pi}{t_0} + s_N\right)t}{\frac{\pi}{t_0} + s_N} - \frac{2\pi}{t_0\left(\frac{\pi^2}{t_0^2} - s_N^2\right)} \right] \\[3ex]
+ \cos s_N t \left[\frac{\sin\left(\frac{\pi}{t_0} - s_N\right)t}{\frac{\pi}{t_0} - s_N} + \frac{\sin\left(\frac{\pi}{t_0} + s_N\right)t}{\frac{\pi}{t_0} + s_N} \right] \quad ; \quad t < t_0 \\[4ex]
\sin s_N t \left[\frac{\cos\left(\frac{\pi}{t_0} - s_N t_0\right)}{\frac{\pi}{t_0} - s_N} + \frac{\cos\left(\frac{\pi}{t_0} + s_N t_0\right)}{\frac{\pi}{t_0} + s_N} - \frac{2\pi}{t_0\left(\frac{\pi^2}{t_0^2} - s_N^2\right)} \right] \\[3ex]
+ \cos s_N t \left[\frac{\sin\left(\frac{\pi}{t_0} - s_N t_0\right)}{\frac{\pi}{t_0} - s_N} + \frac{\sin\left(\frac{\pi}{t_0} + s_N t_0\right)}{\frac{\pi}{t_0} + s_N} \right] \quad ; \quad t \geq t_0
\end{array}
\right\}
\tag{42}
$$

Taking the inverse transformation of (26), (28), and (30), we have from page 6:

$$L^{-1}\{u_1(s)\} = y_M(t)$$
$$L^{-1}\{u_2(s)\} = \theta_Z(t)$$
$$L^{-1}\{u_3(s)\} = \psi^*(t)$$

Writing Eqs. (32) in matrix form:

$$
\begin{bmatrix} P_1 \\ P_2 \\ P_3 \end{bmatrix} = \begin{bmatrix} c_{11} & c_{12} & c_{13} \\ c_{21} & c_{22} & c_{23} \\ c_{31} & c_{32} & c_{33} \end{bmatrix} \begin{bmatrix} f_1(s) \\ f_2(s) \\ f_3(s) \end{bmatrix} + \begin{bmatrix} K_1 \\ K_2 \\ K_3 \end{bmatrix} \dot{\psi}_0^*
$$

and $[e_{ij}] = \begin{bmatrix} 1 & R & 1 \\ 0 & \alpha_{11} & \alpha_{15} \\ 0 & \alpha_{15} & \alpha_{55} \end{bmatrix}$

Then:

$$
y_m(t) = \frac{1}{b_3} \sum_{i=1}^{3} \sum_{j=1}^{3} A_{i_1} L^{-1} \left\{ \frac{e_{ij} f_j(s)}{s^2} \right\} + \sum_{i=1}^{3} A_{i_1} K_i \psi_0^* t
$$

$$
+ \frac{1}{b_3} \sum_{i=1}^{3} \sum_{j=1}^{3} A_{i_2} L^{-1} \left\{ \frac{e_{ij} f_j(s)}{s^2 + s_1^2} \right\} + \sum_{i=1}^{3} \frac{A_{i_2} K_i \psi_0^* \sin s_1 t}{s_1} \qquad (43)
$$

$$
+ \frac{1}{b_3} \sum_{i=1}^{3} \sum_{j=1}^{3} A_{i_3} L^{-1} \left\{ \frac{e_{ij} f_j(s)}{s^2 + s_2^2} \right\} + \sum_{i=1}^{3} \frac{A_{i_3} K_i \psi_0^* \sin s_2 t}{s_2}
$$

$\theta_Z(t) = $ similar with B_{ij} replacing A_{ij} \qquad (44)

$\psi^*(t) = $ similar with C_{ij} replacing A_{ij} \qquad (45)

Restricting the solutions to independently applied forcing functions, equations (43), (44), and (45) simplify to:

Solution 1: $\dot{\psi}^*(t)$ applied in step

$$
\begin{Bmatrix} y_M(t) \\ \theta_Z(t) \\ \psi^*(t) \end{Bmatrix} = \frac{\psi_0^*}{b_3} \left[t \sum_{i=1}^{3} K_i \begin{Bmatrix} A_{i_1} \\ B_{i_1} \\ C_{i_1} \end{Bmatrix} + \frac{1}{s_1} \sin s_1 t \sum_{i=1}^{3} K_i \begin{Bmatrix} A_{i_2} \\ B_{i_2} \\ C_{i_2} \end{Bmatrix} + \frac{1}{s_2} \sin s_2 t \sum_{i=1}^{3} K_i \begin{Bmatrix} A_{i_3} \\ B_{i_3} \\ C_{i_3} \end{Bmatrix} \right] \qquad (46)
$$

Solution 2(a): T_Z^*, T_Z, F_y applied in square step

$$
\begin{Bmatrix} y_M(t) \\ \theta_Z(t) \\ \psi^*(t) \end{Bmatrix} = \frac{1}{b_3} \begin{bmatrix} T_{Z_{Max}}^* \\ F_{y_{Max}} \\ T_{Z_{Max}} \end{bmatrix} (1) \begin{Bmatrix} t^2/2, & t < t_0 \\ & or \\ tt_0 - t_0^2/2, & t \geq t_0 \end{Bmatrix} \sum_{i=1}^{3} \begin{Bmatrix} A_{i_1} \\ B_{i_1} \\ C_{i_1} \end{Bmatrix} \begin{bmatrix} e_{i_1} \\ e_{i_2} \\ e_{i_3} \end{bmatrix}
$$

23

$$+\frac{1}{s_1^2}\left[(2)\left\{\begin{array}{ll}[1-\cos s_1 t], & t<t_0\\ \quad\quad or\\ [\cos s_1(t-t_0)-\cos s_1 t, & t\geq t_0\end{array}\right.\sum_{i=1}^{3}\left\{\begin{array}{c}A_{i_2}\\B_{i_2}\\C_{i_2}\end{array}\right\}\left[\begin{array}{c}e_{i_1}\\e_{i_2}\\e_{i_3}\end{array}\right]\right.$$

$$+\frac{1}{s_2^2}\left[(3)\left\{\begin{array}{ll}[1-\cos s_2 t], & t<t_0\\ \quad\quad or\\ [\cos s_2(t-t_0)-\cos s_2 t, & t\geq t_0\end{array}\right.\sum_{i=1}^{3}\left\{\begin{array}{c}A_{i_3}\\B_{i_3}\\C_{i_3}\end{array}\right\}\left[\begin{array}{c}e_{i_1}\\e_{i_2}\\e_{i_3}\end{array}\right]\right.$$

(47)

- Note: $(N)\left\{\begin{array}{c}\\ \\ \end{array}\right\}$ Terms of this type will be referred to later in the analysis

Solution 2(b): T_Z^*, T_Z, F_y applied in ramp

$$\left\{\begin{array}{c}y_M(t)\\\theta_Z(t)\\\psi^*(t)\end{array}\right\}=\frac{1}{b^3}\left[\begin{array}{c}T_{Z_{Max}}^*\\F_{y_{Max}}\\T_{Z_{Max}}\end{array}\right]\left[(4)\left\{\begin{array}{ll}t^3/6t_0, & t<t_0\\ \quad\quad or\\ \dfrac{tt_0}{2}-\dfrac{t_0^2}{3}, & t\geq t_0\end{array}\right.\sum_{i=1}^{3}\left\{\begin{array}{c}A_{i_1}\\B_{i_1}\\C_{i_1}\end{array}\right\}\left[\begin{array}{c}e_{i_1}\\e_{i_2}\\e_{i_3}\end{array}\right]\right.$$

$$+\frac{1}{s_1^3 t_0}\left\{\begin{array}{ll}s_1 t-\sin s_1 t, & t<t_0\\ \quad\quad or\\ s_1 t\cos s_1(t-t_0)-\sin s_1 t-s_1(t-t_0)\cos s_1(t-t_0)+\sin s_1(t-t_0), & t\geq t_0\end{array}\right.\sum_{i=1}^{3}\left\{\begin{array}{c}A_{i_2}\\B_{i_2}\\C_{i_2}\end{array}\right\}\left[\begin{array}{c}e_{i_1}\\e_{i_2}\\e_{i_3}\end{array}\right]$$

$$+\frac{1}{s_2^3 t_0}\left\{\begin{array}{ll}s_2 t-\sin s_2 t, & t<t_0\\ \quad\quad or\\ s_2 t\cos s_2(t-t_0)-\sin s_2 t-s_2(t-t_0)\cos s_2(t-t_0)+\sin s_2(t-t_0), & t\geq t_0\end{array}\right.\sum_{i=1}^{3}\left\{\begin{array}{c}A_{i_3}\\B_{i_3}\\C_{i_3}\end{array}\right\}\left[\begin{array}{c}e_{i_1}\\e_{i_2}\\e_{i_3}\end{array}\right]$$

(48)

Solution 2(c): T_Z^*, T_Z, F_y applied in half sine

$$\left\{\begin{array}{c}y_M(t)\\\theta_Z(t)\\\psi^*(t)\end{array}\right\}=\frac{1}{b^3}\left[\begin{array}{c}T_{Z_{Max}}^*\\F_{y_{Max}}\\T_{Z_{Max}}\end{array}\right]\left[(7)\left\{\begin{array}{ll}\dfrac{t_0 t}{\pi}\left[1-\cos\dfrac{\pi t}{t_0}\right]-\dfrac{t_0^2}{\pi^2}\left[\sin\dfrac{\pi t}{t_0}+\dfrac{\pi t}{t_0}\cos\dfrac{\pi t}{t_0}\right], & t<t_0\\ \quad\quad or\\ \dfrac{2tt_0}{\pi}-\dfrac{t_0^2}{\pi}, & t\geq t_0\end{array}\right.\sum_{i=1}^{3}\left\{\begin{array}{c}A_{i_1}\\B_{i_1}\\C_{i_1}\end{array}\right\}\left[\begin{array}{c}e_{i_1}\\e_{i_2}\\e_{i_3}\end{array}\right]\right.$$

$$-\frac{1}{2s_1}\left\{\begin{array}{l}\sin s_1 t\left[\dfrac{\cos\left(\dfrac{\pi}{t_0}-s_1\right)t}{\pi/t_0-s_1}+\dfrac{\cos\left(\dfrac{\pi}{t_0}+s_1\right)t}{\pi/t_0+s_1}-\dfrac{2\pi}{t_0\left(\dfrac{\pi^2}{t_0^2}-s_1^2\right)}\right]+\cos s_1 t\left[\dfrac{\sin\left(\dfrac{\pi}{t_0}-s_1\right)t}{\pi/t_0-s_1}-\dfrac{\sin\left(\dfrac{\pi}{t_0}+s_1\right)t}{\pi/t_0+s_1}\right],\quad t<t_0\\[20pt] \qquad\qquad\qquad\qquad\qquad or \\[10pt] \sin s_1 t\left[\dfrac{\cos(\pi-s_1 t_0)}{\pi/t_0-s_1}+\dfrac{\cos(\pi+s_1 t_0)}{\pi/t_0+s_1}-\dfrac{2\pi}{t_0\left(\dfrac{\pi^2}{t_0^2}-s_1^2\right)}\right]+\cos s_1 t\left[\dfrac{\sin(\pi-s_1 t_0)}{\pi/t_0-s_1}-\dfrac{\sin(\pi+s_1 t_0)}{\pi/t_0+s_1}\right],\quad t\ge t_0\end{array}\right\}^{(8)}$$

$$\bullet\ \sum_{i=1}^{3}\begin{Bmatrix}A_{i_2}\\B_{i_2}\\C_{i_2}\end{Bmatrix}\begin{bmatrix}e_{i_1}\\e_{i_2}\\e_{i_3}\end{bmatrix}$$

$$-\frac{1}{2s_2}\left\{\begin{array}{l}\sin s_2 t\left[\dfrac{\cos\left(\dfrac{\pi}{t_0}-s_2\right)t}{\pi/t_0-s_2}+\dfrac{\cos\left(\dfrac{\pi}{t_0}+s_2\right)t}{\pi/t_0+s_2}-\dfrac{2\pi}{t_0\left(\dfrac{\pi^2}{t_0^2}-s_2^2\right)}\right]+\cos s_2 t\left[\dfrac{\sin\left(\dfrac{\pi}{t_0}-s_2\right)t}{\pi/t_0-s_2}-\dfrac{\sin\left(\dfrac{\pi}{t_0}+s_2\right)t}{\pi/t_0+s_2}\right],\quad t<t_0\\[20pt] \qquad\qquad\qquad\qquad\qquad or \\[10pt] \sin s_2 t\left[\dfrac{\cos(\pi-s_2 t_0)}{\pi/t_0-s_2}+\dfrac{\cos(\pi+s_2 t_0)}{\pi/t_0+s_2}-\dfrac{2\pi}{t_0\left(\dfrac{\pi^2}{t_0^2}-s_2^2\right)}\right]+\cos s_2 t\left[\dfrac{\sin(\pi-s_2 t_0)}{\pi/t_0-s_2}-\dfrac{\sin(\pi+s_2 t_0)}{\pi/t_0+s_2}\right],\quad t\ge t_0\end{array}\right\}^{(9)}\qquad (49)$$

$$\bullet\ \sum_{i=1}^{3}\begin{Bmatrix}A_{i_3}\\B_{i_3}\\C_{i_3}\end{Bmatrix}\begin{bmatrix}e_{i_1}\\e_{i_2}\\e_{i_3}\end{bmatrix}$$

2.3.3   Relative Displacements

The displacements in the form given by Eqs. (46) thru (49) are not sufficiently descriptive of system response, since they represent absolute displacements relative to the initial position. To obtain displacements relative to the rigid body displacements, we have from appendix 1:

$$y = y' - \ell\theta \qquad\qquad \ell = R$$
$$\phi = \phi' - \theta$$

Where
$$y' \longrightarrow y_M$$
$$\theta' \longrightarrow \theta_Z$$
$$\theta \longrightarrow \psi^*$$

25

and y = deflection relative to rigid body displacement, θ = angular rotation relative rigid body rotation

Thus, $\begin{Bmatrix} y = y_M - R\psi^* \\ \phi = \theta_Z - \psi^* \end{Bmatrix}$

Now, since the only difference in the solutions given by Eqs. (46) thru (49) for $y_M(t)$, $\theta_Z(t)$, and $\psi^*(t)$ are those caused by the terms A_{ij}, B_{ij}, and C_{ij}. The solutions for the relative displacements can be written immediately.

Let $\begin{Bmatrix} D_{ij} = A_{ij} - RC_{ij} \\ E_{ij} = B_{ij} - C_{ij} \end{Bmatrix}$

Solution 1

$$\begin{Bmatrix} y \\ \phi \end{Bmatrix} = \frac{\ddot{\psi}_0^*}{b^3}\left[t\sum_{i=1}^{3} K_i\begin{Bmatrix} D_{i_1} \\ E_{i_1} \end{Bmatrix} + \frac{1}{s_1}\sin s_1 t\sum_{i=1}^{3} K_i\begin{Bmatrix} D_{i_2} \\ E_{i_2} \end{Bmatrix} + \frac{1}{s_2}\sin s_2 t\sum_{i=1}^{3} K_i\begin{Bmatrix} D_{i_3} \\ E_{i_3} \end{Bmatrix} \right] \tag{50}$$

Solution 2(a)

$$\begin{Bmatrix} y \\ \phi \end{Bmatrix} = \frac{1}{b^3}\begin{bmatrix} T_{Z_{Max}}^* \\ F_{Y_{Max}} \\ T_{Z_{Max}} \end{bmatrix} \left[{}^{(1)}\left\{ \sum_{i=1}^{3}\begin{Bmatrix} D_{i_1} \\ E_{i_1} \end{Bmatrix}\begin{bmatrix} e_{i_1} \\ e_{i_2} \\ e_{i_3} \end{bmatrix} \right\} + \frac{1}{s_1^2}{}^{(2)}\left\{ \sum_{i=1}^{3}\begin{Bmatrix} D_{i_2} \\ E_{i_2} \end{Bmatrix}\begin{bmatrix} e_{i_1} \\ e_{i_2} \\ e_{i_3} \end{bmatrix} \right\} + \frac{1}{s_2^2}{}^{(3)}\left\{ \sum_{i=1}^{3}\begin{Bmatrix} D_{i_3} \\ E_{i_3} \end{Bmatrix}\begin{bmatrix} e_{i_1} \\ e_{i_2} \\ e_{i_3} \end{bmatrix} \right\} \right] \tag{51}$$

Solution 2(b)

$$\begin{Bmatrix} y \\ \phi \end{Bmatrix} = \frac{1}{b^3}\begin{bmatrix} T_{Z_{Max}}^* \\ F_{Y_{Max}} \\ T_{Z_{Max}} \end{bmatrix} \left[{}^{(4)}\left\{ \sum_{i=1}^{3}\begin{Bmatrix} D_{i_1} \\ E_{i_1} \end{Bmatrix}\begin{bmatrix} e_{i_1} \\ e_{i_2} \\ e_{i_3} \end{bmatrix} \right\} + \frac{1}{s_1^2}{}^{(5)}\left\{ \sum_{i=1}^{3}\begin{Bmatrix} D_{i_2} \\ E_{i_2} \end{Bmatrix}\begin{bmatrix} e_{i_1} \\ e_{i_2} \\ e_{i_3} \end{bmatrix} \right\} + \frac{1}{s_2^2}{}^{(6)}\left\{ \sum_{i=1}^{3}\begin{Bmatrix} D_{i_3} \\ E_{i_3} \end{Bmatrix}\begin{bmatrix} e_{i_1} \\ e_{i_2} \\ e_{i_3} \end{bmatrix} \right\} \right] \tag{52}$$

Solution 2(c)

$$\begin{Bmatrix} y \\ \phi \end{Bmatrix} = \frac{1}{b^3}\begin{bmatrix} T_{Z_{Max}}^* \\ F_{Y_{Max}} \\ T_{Z_{Max}} \end{bmatrix} \left[{}^{(7)}\left\{ \sum_{i=1}^{3}\begin{Bmatrix} D_{i_1} \\ E_{i_1} \end{Bmatrix}\begin{bmatrix} e_{i_1} \\ e_{i_2} \\ e_{i_3} \end{bmatrix} \right\} + \frac{1}{s_1^2}{}^{(8)}\left\{ \sum_{i=1}^{3}\begin{Bmatrix} D_{i_2} \\ E_{i_2} \end{Bmatrix}\begin{bmatrix} e_{i_1} \\ e_{i_2} \\ e_{i_3} \end{bmatrix} \right\} + \frac{1}{s_2^2}{}^{(9)}\left\{ \sum_{i=1}^{3}\begin{Bmatrix} D_{i_3} \\ E_{i_3} \end{Bmatrix}\begin{bmatrix} e_{i_1} \\ e_{i_2} \\ e_{i_3} \end{bmatrix} \right\} \right] \tag{53}$$

In these solutions, the terms denoted thus ${}^{(N)}\{\ \}$, refer to terms in the solutions of Eqs. (46) thru (49).

2.3.4 Discussion:

Equations (50) thru (53) above give the transient response to the specified forcing functions in final form. From these equations, forces at the interface of the arm and the gimbal ring may be calculated. However, the main interest of this study in in the calculation of dynamic response factors, not in the interface forces. Thus, we will now concern ourselves with this topic.

2.3.5 Dynamic Response Factors

The dynamic response factor "γ" is defined as the ratio of the maximum transient response to the static displacement under the peak value of transient force. Since solution 1 does not involve the input forces, it is not useful in this calculation.

The maximum values of the relative displacements can be obtained by equating the time derivatives of Eqs. (51) thru (53) to zero, and substituting the values of time which satisfy these equations back into (51) thru (53).

Solution 2(a): Maximizing Eq. (51)

$$
\left\{ \begin{matrix} 0 \\ 0 \end{matrix} \right\} = \left\{ \begin{matrix} t, & t < t_0 \\ or \\ t_0, & t \ge t_0 \end{matrix} \right\} \sum_{i=1}^{3} \left\{ \begin{matrix} D_{i_1} \\ E_{i_1} \end{matrix} \right\} \left[\begin{matrix} e_{i_1} \\ e_{i_2} \\ e_{i_3} \end{matrix} \right] + \frac{1}{s_1^2} \left\{ \begin{matrix} s_1 \sin s_1 t, & t < t_0 \\ or \\ -s_1 \sin s_1 (t - t_0) + s_1 \sin s_1 t, & t \ge t_0 \end{matrix} \right\} \sum_{i=1}^{3} \left\{ \begin{matrix} D_{i_2} \\ E_{i_2} \end{matrix} \right\} \left[\begin{matrix} e_{i_1} \\ e_{i_2} \\ e_{i_3} \end{matrix} \right]
$$
$$
+ \frac{1}{s_2^2} \left\{ \begin{matrix} s_2 \sin s_2 t, & t < t_0 \\ or \\ -s_2 \sin s_2 (t - t_0) + s_2 \sin s_2 t, & t \ge t_0 \end{matrix} \right\} \sum_{i=1}^{3} \left\{ \begin{matrix} D_{i_3} \\ E_{i_3} \end{matrix} \right\} \left[\begin{matrix} e_{i_1} \\ e_{i_2} \\ e_{i_3} \end{matrix} \right] \tag{51}
$$

Denoting the values of t which satisfy the Eqs. (51) for t<t_0 by $\left\{ \begin{matrix} t_{1_\gamma} \\ t_{1_\theta} \end{matrix} \right\}$, and the values of t which

satisfy it for $t \ge t_0$ by $\left\{ \begin{matrix} t_{2_\gamma} \\ t_{2_\theta} \end{matrix} \right\}$, providing such values exist, we substitute these values into Eq. (51) and

obtain $\left\{ \begin{matrix} y_{max} \\ \theta_{Max} \end{matrix} \right\}$ for unit values of the peak forcing functions. This is the maximum transient response vector.

The static displacements which would occur under the peak values of the transient forces (equal to unity, here) may be obtained as follows.

The inertia forces acting on the arm as a result of $T^*_{Z_{Max}}$ are obtained by considering only the rigid

body response. Hence, $$\ddot{\psi}^* = \frac{T^*_{z_{Max}}}{I_{Z_{Tot}}} = \frac{T^*_{z_{Max}}}{K_1}$$

Inertia Forces Acting on the Arm

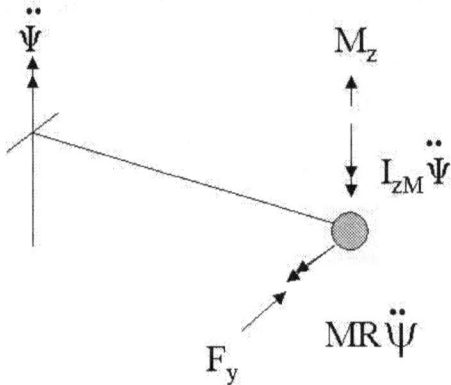

FIGURE 2.3.5

Thus, $$\left\{ \begin{array}{l} P_{Y_{T_Z^*}} = -MR\,\ddot{\psi}^* = -MR\dfrac{T^*_{Z_{Max}}}{K_1} \\[2em] M_{Z_{T_Z^*}} = -I_{Z_{Max}}\,\ddot{\psi}^* = -I_{Z_{Max}}\dfrac{T^*_{Z_{Max}}}{K_1} \end{array} \right\}$$

Forces acting on the arm as a result of $F_{Y_{Max}}$ and $T_{Z_{Max}}$ applied at the mass are found in a similar

manner. Thus,

$$\ddot{\psi}^*_{F_Y} = \frac{F_{Y_{Max}} R}{K_1} \quad ; \quad \ddot{\psi}^*_{T_Z} = \frac{T_{Z_{Max}}}{K_1}$$

$$\left\{ \begin{array}{l} P_{Y_{FY}} = -MR\ddot{\psi}^* + F_{Y_{Max}} = \left(1 - \dfrac{MR^2}{K_1}\right)F_{Y_{Max}} \\[4mm] M_{Z_{FY}} = -I_{Z_M}\ddot{\psi}^* = -I_{Z_M}\dfrac{R}{R_1}F_{Y_{Max}} \end{array} \right\}$$

and,

$$\left\{ \begin{array}{l} P_{Y_{TZ}} = MR\dfrac{T_{Z_{Max}}}{K_1} \\[4mm] M_{Z_{TZ}} = \left(1 - \dfrac{I_{Z_M}}{K_1}\right)T_{Z_{Max}} \end{array} \right\}$$

From Appendix 1, page 1:

$$\left\{ \begin{array}{l} S_y = \alpha_{11}P_y + \alpha_{15}M_Z \equiv y_s \\ \theta_Z = \alpha_{51}P_y + \alpha_{55}M_Z \equiv \theta_s \end{array} \right\}$$

Hence, the static deflections $\left\{ \begin{array}{c} y_s \\ \theta_s \end{array} \right\}$ may be obtained for unit values of the peak transient forces:

$$j = 1: \quad T_{Z_{Max}}^* = 1 \quad \left\{ \begin{array}{l} y_s = -\alpha_{11}M\,R/K_1 - \alpha_{15}\,I_{Z_M}/K_1 \\ \phi_s = -\alpha_{51}M\,R/K_1 - \alpha_{55}\,I_{Z_M}/K_1 \end{array} \right\}$$

$$j = 2: \quad F_{Y_{Max}} = 1 \quad \left\{ \begin{array}{l} y_s = \alpha_{11}\left(1 - M\,R^2/K_1\right) - \alpha_{15}\,I_{Z_M}R/K_1 \\ \phi_s = \alpha_{51}\left(1 - M\,R^2/K_1\right) - \alpha_{55}\,I_{Z_M}R/K_1 \end{array} \right\}$$

$$j = 3: \quad T_{Z_{Max}} = 1 \quad \left\{ \begin{array}{l} y_s = -\alpha_{11}M\,R/K_1 + \alpha_{15}\left(1 - I_{Z_{Max}}/K_1\right) \\ \phi_s = -\alpha_{51}M\,R/K_1 + \alpha_{55}\left(1 - I_{Z_{Max}}/K_1\right) \end{array} \right\}$$

Having these static deflections available, the dynamic response factors may be calculated for each of the forcing functions and for each of the displacements, θ_s and y_s. Thus,

<u>Square step</u> $\left\{ \begin{array}{ll} \gamma_{j_y}^{\Diamond} = y_{j_{Max}}^{\Diamond}/y_s \quad , & j = 1, 2, 3 \\ \gamma_{j_\theta}^{\Diamond} = \theta_{j_{Max}}^{\Diamond}/\theta_s \quad , & j = 1, 2, 3 \end{array} \right\}$

Solution 2(b): Performing the maximization, we have:

$$\left\{ \begin{array}{c} 0 \\ 0 \end{array} \right\} = \left\{ \begin{array}{c} \dfrac{t^2}{2t_0}, t < t_0 \\ or \\ \dfrac{t_0}{2}, t \ge 0 \end{array} \right\} \sum_{i=1}^{3} \left\{ \begin{array}{c} D_{i_1} \\ E_{i_1} \end{array} \right\} \left[\begin{array}{c} e_{i_1} \\ e_{i_2} \\ e_{i_3} \end{array} \right]$$

$$+ \dfrac{1}{s_1^3 t_0} \left\{ \begin{array}{c} s_1 - s_1 \cos s_1 t, \quad t < t_0 \\ or \\ s_1 \cos s_1(t - t_0) - s_1 \cos s_1 t - s_1^2 t_0 \sin s_1(t - t_0), t \ge 0 \end{array} \right\} \cdot \sum_{i=1}^{3} \left\{ \begin{array}{c} D_{i_2} \\ E_{i_2} \end{array} \right\} \left[\begin{array}{c} e_{i_1} \\ e_{i_2} \\ e_{i_3} \end{array} \right]$$

29

$$+\frac{1}{s_2^3 t_0}\left\{\begin{array}{c} s_2 - s_2\cos s_2 t, \quad t < t_0 \\ or \\ s_2\cos s_2(t-t_0) - s_2\cos s_2 t - s_2^2 t_0\sin s_2(t-t_0), t \geq 0 \end{array}\right\} \bullet \sum_{i=1}^{3}\left\{\begin{array}{c} D_{i_3} \\ E_{i_3} \end{array}\right\}\begin{bmatrix} e_{i_1} \\ e_{i_2} \\ e_{i_3} \end{bmatrix} \qquad (52)$$

Following a similar process as outlined for 2(a), we obtain:

$$\underline{\text{Ramp}}\left\{\begin{array}{l} \gamma_{j_y}^{\Delta} = {y_{Max}^{\Delta^j}}\Big/{y_s} \\ \\ \gamma_{j_\phi}^{\Delta} = {\phi_{Max}^{\Delta^j}}\Big/{\phi_s} \end{array}\right. \quad ; \quad j = 1, 2, 3 \right\}$$

Solutions 2(c)

$$\left\{\begin{array}{c} 0 \\ 0 \end{array}\right\} = \left\{\begin{array}{c} \dfrac{t_0}{\pi}\left[1 - 3\cos\dfrac{\pi t}{t_0}\right] + 2t\sin\dfrac{\pi t}{t_0}, t < t_0 \\ or \\ \dfrac{2t_0}{\pi}, \quad t \geq 0 \end{array}\right\} \sum_{i=1}^{3}\left\{\begin{array}{c} D_{i_1} \\ E_{i_1} \end{array}\right\}\begin{bmatrix} e_{i_1} \\ e_{i_2} \\ e_{i_3} \end{bmatrix}$$

$$-\frac{1}{2s_1}\left\{\begin{array}{c} \cos s_1 t\left[\dfrac{\pi}{\pi-s_1 t_0}\cos\left(\dfrac{\pi}{t_0}-s_1\right)t - \dfrac{\pi}{\pi+s_1 t_0}\cos\left(\dfrac{\pi}{t_0}-s_1\right)t - \dfrac{2\pi s_1}{t_0\left(\dfrac{\pi^2}{t_0^2}-s_1^2\right)}\right] - \sin s_1 t\left\{\dfrac{\pi}{\pi-s_1 t_0}\sin\left(\dfrac{\pi}{t_0}-s_1\right)t + \dfrac{\pi}{\pi+s_1 t_0}\sin\left(\dfrac{\pi}{t_0}+s_1\right)t\right\}, t < t_0 \\ or \\ s_1\cos s_1 t\left[\dfrac{\cos(\pi-s_1 t_0)}{\pi/t_0 - s_1} + \dfrac{\cos(\pi+s_1 t_0)}{\pi/t_0 + s_1} - \dfrac{2\pi}{t_0\left(\dfrac{\pi^2}{t_0^2}-s_1^2\right)}\right] - s_1\sin s_1 t\left[\dfrac{\sin(\pi-s_1 t_0)}{\pi/t_0 - s_1} + \dfrac{\sin(\pi+s_1 t_0)}{\pi/t_0 + s_1}\right], t \geq 0 \end{array}\right\}$$

$$\bullet\sum_{i=1}^{3}\left\{\begin{array}{c} D_{i_2} \\ E_{i_2} \end{array}\right\}\begin{bmatrix} e_{i_1} \\ e_{i_2} \\ e_{i_3} \end{bmatrix} \quad -\frac{1}{2s_2}\left\{\begin{array}{c} \\ \\ \\ S_2 \end{array}\right\} \sum_{i=1}^{3}\left\{\begin{array}{c} D_{i_2} \\ E_{i_2} \end{array}\right\}\begin{bmatrix} e_{i_1} \\ e_{i_2} \\ e_{i_3} \end{bmatrix} \qquad (53)$$

Following the procedure used for 2(a), we obtain:

$$\underline{\text{Half Sine}}\left\{\begin{array}{l} \gamma_{j_y}^{\cap} = {y_{Max}^{\cap^j}}\Big/{y_s} \\ \\ \gamma_{j_\phi}^{\cap} = {\phi_{Max}^{\cap^j}}\Big/{\phi_s} \end{array}\right. \quad ; \quad j = 1, 2, 3 \right\}$$

These arm dynamic response factors are the final result of the analysis for dynamic response of the system in the Y plane.

2.4 Z Axis Analysis

2.4.1 Solution for Z Axis Transient Response

Equations (10) are operated on by the Laplacian operator $s = \dfrac{d}{dt}$,

where $\begin{aligned} L\{Z_M(t)\} &= v_1(s) \\ L\{\theta_Y(t)\} &= v_2(s) \end{aligned}$ $\qquad \begin{aligned} L\{F_Z(t)\} &= g_1(s) \\ L\{T_Y(t)\} &= g_2(s) \end{aligned}$

Assuming the initial conditions:

$$Z_0 = \dot{Z}_0 = \theta_{Y_0} = \dot{\theta}_{Y_0} = 0$$

We have:

$$v_1 + \alpha_{22} M s^2 v_1 + \alpha_{24} I_{Y_M} s^2 v_2 = \alpha_{22} g_1(s) + \alpha_{24} g_2(s)$$

$$v_2 + \alpha_{42} M s^2 v_1 + \alpha_{44} I_{Y_M} s^2 v_2 = \alpha_{42} g_1(s) + \alpha_{44} g_2(s)$$

or

$$(1 + \alpha_{22} M s^2) v_1 + \alpha_{24} I_{Y_M} s^2 v_2 = \alpha_{22} g_1(s) + \alpha_{24} g_2(s) = R_1$$

$$\alpha_{42} M s^2 v_1 + (1 + \alpha_{44} I_{Y_M} s^2) v_2 = \alpha_{42} g_1(s) + \alpha_{44} g_2(s) = R_2 \tag{54}$$

Solving for v_1 and v_2:

$$v_1 = \frac{\begin{vmatrix} R_1 & \alpha_{24} I_{Y_M} s^2 \\ R_2 & 1 + \alpha_{44} I_{Y_M} s^2 \end{vmatrix}}{\Delta} = R_1 \frac{1 + \alpha_{44} I_{Y_M} s^2}{\Delta} + R_2 \frac{-\alpha_{24} I_{Y_M} s^2}{\Delta}$$

$$v_2 = \frac{\begin{vmatrix} 1 + \alpha_{22} M s^2 & R_1 \\ \alpha_{42} M s^2 & R_2 \end{vmatrix}}{\Delta} = R_1 \frac{-\alpha_{42} M s^2}{\Delta} + R_2 \frac{1 + \alpha_{22} M s^2}{\Delta} \tag{55}$$

where the characteristic equation is:

$$\Delta = (1 + \alpha_{22} M s^2)(1 + \alpha_{44} I_{Y_M} s^2) - \alpha_{42} M s^2 \alpha_{24} I_{Y_M} s^2 = 0$$

or

$$(\alpha_{22} M \alpha_{44} I_{Y_M} - \alpha_{42} M \alpha_{24} I_{Y_M}) s^4 + (\alpha_{22} M + \alpha_{44} I_{Y_M}) s^2 + 1 = 0 \tag{56}$$

letting:
$$d_1 = \alpha_{22} \alpha_{44} M I_{Y_M} - \alpha_{42} \alpha_{24} M I_{Y_M}$$
$$d_2 = \alpha_{22} M + \alpha_{44} I_{Y_M}$$

or,
$$d_1 s^4 + d_2 s^2 + 1 = 0 \qquad \text{where:} \qquad \begin{aligned} A &= d_2 / d_1 \\ B &= 1 / d_1 \end{aligned}$$
$$s^4 + A s^2 + B = 0$$

Then: $s^2 = -\dfrac{A}{2} \pm \sqrt{\left(\dfrac{A}{2}\right)^2 - B}$

Denoting $\quad s_1^2 < s_2^2 = (-1)x \quad$ Roots of $\Delta = 0$

We have $\Delta = d_1(s^2 + s_1^2)(s^2 + s_2^2)$; expressed in terms of the characteristic roots.

Hence,

$$d_1 v_1 = R_1 \frac{1 + \alpha_{44} I_{Y_M} s^2}{(s^2 + s_1^2)(s^2 + s_2^2)} + R_2 \frac{-\alpha_{24} I_{Y_M} s^2}{(s^2 + s_1^2)(s^2 + s_2^2)}$$

$$d_1 v_2 = R_1 \frac{-\alpha_{42} M s^2}{(s^2 + s_1^2)(s^2 + s_2^2)} + R_2 \frac{1 + \alpha_{22} M s^2}{(s^2 + s_1^2)(s^2 + s_2^2)}$$

$$(57)$$

Seeking the inverse transforms of (57), we expand the right hand sides using partial fractions.

Let $\quad \begin{array}{ll} a_{11} = \alpha_{44} I_{Y_M} & a_{21} = -\alpha_{42} M \\ a_{12} = -\alpha_{24} I_{Y_M} & a_{22} = \alpha_{22} M \end{array}$

$$d_1 v_1 = R_1 \left\{ \frac{A_{11}}{s^2 + s_1^2} + \frac{A_{12}}{s^2 + s_2^2} \right\} + R_2 \left\{ \frac{A_{21}}{s^2 + s_1^2} + \frac{A_{22}}{s^2 + s_2^2} \right\}$$

$$d_1 v_2 = R_1 \left\{ \frac{B_{11}}{s^2 + s_1^2} + \frac{B_{12}}{s^2 + s_2^2} \right\} + R_2 \left\{ \frac{B_{21}}{s^2 + s_1^2} + \frac{B_{22}}{s^2 + s_2^2} \right\}$$

$$(58)$$

expanding the terms in $\left\{ \quad \right\}$, we have:

$$A_{11}\left(s^2 + s_2^2\right) + A_{12}\left(s^2 + s_1^2\right) = 1 + a_{11}s^2$$
$$A_{21}\left(s^2 + s_2^2\right) + A_{22}\left(s^2 + s_1^2\right) = a_{12}s^2$$
$$B_{11}\left(s^2 + s_2^2\right) + B_{12}\left(s^2 + s_1^2\right) = a_{21}s^2$$
$$B_{21}\left(s^2 + s_2^2\right) + B_{22}\left(s^2 + s_1^2\right) = 1 + a_{22}s^2$$

Setting $s^2 = -s_1^2, \quad -s_2^2$ successively, the following expressions for the coefficients A_{ij}, B_{ij} are obtained:

$$A_{11}(s_2^2 - s_1^2) = 1 - a_{11}s_1^2; \quad A_{11} = \frac{1 - a_{11}s_1^2}{s_2^2 - s_1^2}$$

$$A_{12}(-s_2^2 + s_1^2) = 1 - a_{11}s_2^2; \quad A_{12} = -\frac{1 - a_{11}s_2^2}{s_2^2 - s_1^2}$$

$$A_{21}(s_2^2 - s_1^2) = -a_{12}s_1^2; \quad A_{21} = \frac{-a_{12}s_1^2}{s_2^2 - s_1^2}$$

$$A_{22}(-s_2^2 + s_1^2) = -a_{12}s_2^2; \quad A_{22} = \frac{a_{12}s_2^2}{s_2^2 - s_1^2}$$

$$B_{11}(s_2^2 - s_1^2) = -a_{12}s_1^2; \quad B_{11} = -\frac{a_{21}s_1^2}{s_2^2 - s_1^2}$$

$$B_{12}(-s_2^2 + s_1^2) = -a_{21}s_2^2; \quad B_{12} = \frac{a_{21}s_2^2}{s_2^2 - s_1^2}$$

$$B_{21}(s_2^2 - s_1^2) = 1 - a_{22}s_1^2; \quad B_{21} = \frac{1 - a_{22}s_1^2}{s_2^2 - s_1^2}$$

$$B_{22}(-s_2^2 + s_1^2) = 1 - a_{22}s_2^2; \quad B_{22} = -\frac{1 - a_{22}s_2^2}{s_2^2 - s_1^2}$$

We write the forcing function in matrix notation as:

$$\begin{bmatrix} R_1 \\ R_2 \end{bmatrix} = \begin{bmatrix} e_{11} & e_{12} \\ e_{21} & e_{22} \end{bmatrix} \begin{bmatrix} g_1(s) \\ g_2(s) \end{bmatrix} \tag{59}$$

where, from page 26:

$$e_{11} = \alpha_{22} \quad e_{12} = \alpha_{24}$$
$$e_{21} = \alpha_{42} \quad e_{22} = \alpha_{44}$$

The inverse transforms of (58) will now be formed. We use the convolution theorem, since

$$L^{-1}\{f(s)g(s)\} = \int_0^t F(\beta)G(t - \beta)d\beta$$

Where $\quad L^{-1}\{f(s)\} = F(t), \quad L^{-1}\{g(s)\} = G(t)$

Thus, $L^{-1}\left\{ \dfrac{eg_i(s)}{s^2 + s_N^2} \right\} = \displaystyle\int_0^t \dfrac{e}{s_N} G_i(\beta) \sin s_N(t - \beta)d\beta \quad \begin{cases} i = 1,2 \\ N = 1,2 \end{cases}$

Now, $\quad \begin{aligned} L^{-1}\{g_1(s)\} &= F_Z(t) \\ L^{-1}\{g_2(s)\} &= T_Y(t) \end{aligned} \tag{60}$

We will restrict our investigation to the three types of functions given on pages 14 and 15, Eqs. (38), (40), and (42). Therefore give the inverse transforms of the terms encountered in Eqs. (58).

Now, $\quad \begin{aligned} L^{-1}\{v_1(s)\} &= Z_M(t) \\ L^{-1}\{v_2(s)\} &= \theta_Y(t) \end{aligned} \tag{61}$

The solution for the transient response of the system can be written as:

$$Z_M(t) = \frac{1}{d_1}\sum_{i=1}^{Z}\sum_{j=1}^{Z} A_{i_1} L^{-1}\left\{\frac{e_{ij}g_j(s)}{s^2+s_1^2}\right\} + \frac{1}{d_1}\sum_{i=1}^{Z}\sum_{j=1}^{Z} A_{i_2} L^{-1}\left\{\frac{e_{ij}g_j(s)}{s^2+s_2^2}\right\}$$

$$\theta_Y(t) = \frac{1}{d_1}\sum_{i=1}^{Z}\sum_{j=1}^{Z} B_{i_1} L^{-1}\left\{\frac{e_{ij}g_j(s)}{s^2+s_1^2}\right\} + \frac{1}{d_1}\sum_{i=1}^{Z}\sum_{j=1}^{Z} B_{i_2} L^{-1}\left\{\frac{e_{ij}g_j(s)}{s^2+s_2^2}\right\}$$

(62)

Restricting the solution to independently applied forcing functions, equations (62) simplify as follows:

$$\left\{\begin{matrix}Z_M(t)\\\theta_Y(t)\end{matrix}\right\} = \frac{1}{d_1}\sum_{i=1}^{Z}\left\{\begin{matrix}A_{i_1}\\B_{i_1}\end{matrix}\right\} L^{-1}\left[\left\{\begin{matrix}\dfrac{e_{i_1}g_1(s)}{s^2+s_1^2}\\[2mm]\dfrac{e_{i_2}g_2(s)}{s^2+s_1^2}\end{matrix}\right\}\right] + \frac{1}{d_1}\sum_{i=1}^{Z}\left\{\begin{matrix}A_{i_2}\\B_{i_2}\end{matrix}\right\} L^{-1}\left[\left\{\begin{matrix}\dfrac{e_{i_1}g_1(s)}{s^2+s_2^2}\\[2mm]\dfrac{e_{i_2}g_2(s)}{s^2+s_2^2}\end{matrix}\right\}\right]$$

(63)

Where either $g_1(s)$ or $g_2(s)$ is applied independently.

Solution (a): F_Z or T_Y applied in square step.

$$\left\{\begin{matrix}Z_M(t)\\\theta_Y(t)\end{matrix}\right\} = \frac{1}{d_1}\left[\begin{matrix}F_{Z_{Max}}\\T_{Y_{Max}}\end{matrix}\right]\left[\left(\left\{\begin{matrix}\dfrac{1}{s_1^2}[1-\cos s_1 t], & t<t_0\\ or\\ \dfrac{1}{s_1^2}[\cos s_1(t-t_0)-\cos s_1 t], & t\geq t_0\end{matrix}\right\}\sum_{i=1}^{2}\left[\begin{matrix}e_{i_1}\\e_{i_2}\end{matrix}\right]\left\{\begin{matrix}A_{i_1}\\B_{i_1}\end{matrix}\right\}\right) + \left(\left\{\begin{matrix}\dfrac{1}{s_2^2}[1-\cos s_2 t], & t<t_0\\ or\\ \dfrac{1}{s_2^2}[\cos s_2(t-t_0)-\cos s_2 t], & t\geq t_0\end{matrix}\right\}\sum_{i=1}^{2}\left[\begin{matrix}e_{i_1}\\e_{i_2}\end{matrix}\right]\left\{\begin{matrix}A_{i_2}\\B_{i_2}\end{matrix}\right\}\right)\right]$$

(64)

Solution (b): F_Z or T_Y applied in ramp.

$$\begin{Bmatrix} Z_M(t) \\ \theta_Y(t) \end{Bmatrix} = \frac{1}{d_1} \begin{bmatrix} F_{Z_{Max}} \\ T_{Y_{Max}} \end{bmatrix} \left[\begin{Bmatrix} s_1 t - \sin s_1 t, \quad t < t_0 \\ or \\ s_1 t \cos s_1(t-t_0) - \sin s_1 t - s_1(t-t_0)\cos s_1(t-t_0) + \sin s_1(t-t_0), t \geq t_0 \end{Bmatrix} \right.$$
$$\bullet \frac{1}{s_1^3 t_0} \sum_{i=1}^{2} \begin{bmatrix} e_{i_1} \\ e_{i_2} \end{bmatrix} \begin{Bmatrix} A_{i_1} \\ B_{i_1} \end{Bmatrix}$$
$$+ \begin{Bmatrix} s_2 t - \sin s_2 t, \quad t < t_0 \\ or \\ s_2 t \cos s_2(t-t_0) - \sin s_2 t - s_2(t-t_0)\cos s_2(t-t_0) + \sin s_2(t-t_0), t \geq t_0 \end{Bmatrix}$$
$$\left. \bullet \frac{1}{s_2^3 t_0} \sum_{i=1}^{2} \begin{bmatrix} e_{i_1} \\ e_{i_2} \end{bmatrix} \begin{Bmatrix} A_{i_2} \\ B_{i_2} \end{Bmatrix} \right]$$

(65)

Solution (c): F_Z or T_Y applied in half sine.

$$\begin{Bmatrix} Z_M(t) \\ \theta_Y(t) \end{Bmatrix} = \frac{-1}{d_1} \begin{bmatrix} F_{Z_{Max}} \\ T_{Y_{Max}} \end{bmatrix} \left[\begin{Bmatrix} \sin s_1 t \left[\dfrac{\cos\left(\dfrac{\pi}{t_0} - s_1\right)t}{\dfrac{\pi}{t_0} - s_1} + \dfrac{\cos\left(\dfrac{\pi}{t_0} + s_1\right)t}{\dfrac{\pi}{t_0} + s_1} - \dfrac{2\pi}{t_0\left(\dfrac{\pi^2}{t_0^2} - s_1^2\right)} \right] \\ + \cos s_1 t \left[\dfrac{\sin\left(\dfrac{\pi}{t_0} - s_1\right)t}{\dfrac{\pi}{t_0} - s_1} - \dfrac{\sin\left(\dfrac{\pi}{t_0} + s_1\right)t}{\dfrac{\pi}{t_0} + s_1} \right], \quad t < t_0 \end{Bmatrix} \right.$$
$$or$$
$$\left. + \begin{Bmatrix} \sin s_1 t \left[\dfrac{\cos(\pi - t_0 s_1)}{\dfrac{\pi}{t_0} - s_1} + \dfrac{\cos(\pi + t_0 s_1)}{\dfrac{\pi}{t_0} + s_1} - \dfrac{2\pi}{t_0\left(\dfrac{\pi^2}{t_0^2} - s_1^2\right)} \right] \\ + \cos s_1 t \left[\dfrac{\sin(\pi - t_0 s_1)}{\dfrac{\pi}{t_0} - s_1} - \dfrac{\sin(\pi - t_0 s_1)}{\dfrac{\pi}{t_0} + s_1} \right], \quad t \geq t_0 \end{Bmatrix} \frac{1}{2s_1} \sum_{i=1}^{2} \begin{bmatrix} e_{i_1} \\ e_{i_2} \end{bmatrix} \begin{Bmatrix} A_{i_1} \\ B_{i_1} \end{Bmatrix} \right]$$

35

$$
+ \left\{ \left[\begin{array}{l} \left[\sin s_2 t \left[\dfrac{\cos\left(\dfrac{\pi}{t_0} - s_2\right)t}{\dfrac{\pi}{t_0} - s_2} + \dfrac{\cos\left(\dfrac{\pi}{t_0} + s_2\right)t}{\dfrac{\pi}{t_0} + s_2} - \dfrac{2\pi}{t_0\left(\dfrac{\pi^2}{t_0^2} - s_2^2\right)} \right] \right. \\[3em] \left. + \cos s_2 t \left[\dfrac{\sin\left(\dfrac{\pi}{t_0} - s_2\right)t}{\dfrac{\pi}{t_0} - s_2} - \dfrac{\sin\left(\dfrac{\pi}{t_0} + s_2\right)t}{\dfrac{\pi}{t_0} + s_2} \right] \right], \quad t < t_0 \\[4em] or \\[2em] \left[\sin s_2 t \left[\dfrac{\cos(\pi - t_0 s_2)}{\dfrac{\pi}{t_0} - s_2} + \dfrac{\cos(\pi + t_0 s_2)}{\dfrac{\pi}{t_0} + s_2} - \dfrac{2\pi}{t_0\left(\dfrac{\pi^2}{t_0^2} - s_2^2\right)} \right] \right. \\[3em] \left. + \cos s_2 t \left[\dfrac{\sin(\pi - t_0 s_2)}{\dfrac{\pi}{t_0} - s_2} - \dfrac{\sin(\pi - t_0 s_2)}{\dfrac{\pi}{t_0} + s_2} \right] \right], \quad t \geq t_0 \end{array} \right] \right\} \dfrac{1}{2 s_2} \sum_{i=1}^{2} \begin{bmatrix} e_{i_1} \\ e_{i_2} \end{bmatrix} \begin{Bmatrix} A_{i_2} \\ B_{i_2} \end{Bmatrix}
\tag{66}
$$

2.4.2 Dynamic response factors

The static deflections under the peak transient loads $F_{Z_{Max}}, T_{Y_{Max}}$ are from Eq. (10):

$$Z_{M_S} = \alpha_{22} F_{Z_{Max}} + \alpha_{24} T_{Y_{Max}}$$
$$\theta_{Y_S} = \alpha_{42} F_{Z_{Max}} + \alpha_{44} T_{Y_{Max}}$$

Thus, for unit values of the peak loads, we obtain

$$j = 1: \quad F_{Z_{Max}} = 1 \quad \begin{cases} Z_{M_{S(1)}} = \alpha_{22} \\ \theta_{Y_{S(1)}} = \alpha_{42} \end{cases}$$

$$j = 2: \quad T_{Y_{Max}} = 1 \quad \begin{cases} Z_{M_{S(2)}} = \alpha_{24} \\ \theta_{Y_{S(2)}} = \alpha_{44} \end{cases}$$

Knowing the static displacements. The dynamic response factors may be calculated for each of the forcing functions and for each of the displacements Z_{M_S}, θ_{Y_S}.

Solution (a):

Maximizing Eq. (64)

$$\begin{Bmatrix} 0 \\ 0 \end{Bmatrix} = \frac{1}{s_1^2} \begin{cases} s_1 \sin s_1 t, & t < t_0 \\ & or \\ -s_1 \sin s_1 (t - t_0) + s_1 \sin s_1 t, & t \ge t_0 \end{cases} \sum_{i=1}^{2} \begin{bmatrix} A_{i_1} \\ B_{i_1} \end{bmatrix} \begin{bmatrix} e_{i_1} \\ e_{i_2} \end{bmatrix}$$

$$+ \frac{1}{s_2^2} \begin{cases} s_2 \sin s_2 t, & t < t_0 \\ & or \\ -s_2 \sin s_2 (t - t_0) + s_2 \sin s_2 t, & t \ge t_0 \end{cases} \sum_{i=1}^{2} \begin{bmatrix} A_{i_2} \\ B_{i_2} \end{bmatrix} \begin{bmatrix} e_{i_1} \\ e_{i_2} \end{bmatrix} \tag{64'}$$

Denoting the values of t that satisfy the Eqs. For $t < t_0$ by $\begin{Bmatrix} t_{1_z} \\ t_{1_\theta} \end{Bmatrix}$ and for $t \ge t_0$ by $\begin{Bmatrix} t_{2_z} \\ t_{2_\theta} \end{Bmatrix}$, providing

such values exist, we substitute these values into Eq. (64) to obtain $\begin{Bmatrix} Z_{Max}^\Diamond \\ \theta_{Max}^\Diamond \end{Bmatrix}$ for unit values of

$F_{Z_{Max}}, T_{Y_{Max}}$.

The dynamic response factors are, then:

Square step $\begin{Bmatrix} \gamma_{j_z}^\Diamond = Z_{j_{Max}}^\Diamond \Big/ Z_{M_S} \\ \gamma_{j_\theta}^\Diamond = \theta_{j_{Max}}^\Diamond \Big/ \theta_{Y_S} \end{Bmatrix}; \qquad j = 1,2$

Solution (b):

Maximizing Eq. (65):

$$\begin{Bmatrix} 0 \\ 0 \end{Bmatrix} = -\frac{1}{2 s_1^3 t_0} \begin{cases} s_1 - s_1 \cos s_1 t, & t < t_0 \\ & or \\ s_1 \cos s_1 (t - t_0) - s_1 \cos s_1 t - s_1^2 t_0 \sin s_1 (t - t_0), & t \ge t_0 \end{cases} \sum_{i=1}^{2} \begin{bmatrix} A_{i_1} \\ B_{i_1} \end{bmatrix} \begin{bmatrix} e_{i_1} \\ e_{i_2} \end{bmatrix}$$

$$+ \frac{1}{2 s_2^3 t_0} \begin{cases} s_2 - s_2 \cos s_2 t, & t < t_0 \\ & or \\ s_2 \cos s_2 (t - t_0) - s_2 \cos s_2 t - s_2^2 t_0 \sin s_2 (t - t_0), & t \ge t_0 \end{cases} \sum_{i=1}^{2} \begin{bmatrix} A_{i_2} \\ B_{i_2} \end{bmatrix} \begin{bmatrix} e_{i_1} \\ e_{i_2} \end{bmatrix} \tag{65'}$$

Following the procedure outlined above, we obtain

Ramp $\left\{\begin{array}{c}\overset{\Delta}{\gamma}_{j_Z} = \overset{\Delta}{Z}_{j_{Max}} \Big/ Z_{M_S} \\ \overset{\Delta}{\gamma}_{j_\theta} = \overset{\Delta}{\theta}_{j_{Max}} \Big/ \theta_{Y_S}\end{array}\right\}; \qquad j = 1,2$

Solution (c):

Maximizing Eq. (66):

$$\begin{Bmatrix}0\\0\end{Bmatrix} = -\frac{1}{2s_1}\left\{\begin{array}{l}\cos s_1 t\left[\dfrac{\pi}{\pi - s_1 t_0}\cos\left(\dfrac{\pi}{t_0} - s_1\right)t - \dfrac{\pi}{\pi + s_1 t_0}\cos\left(\dfrac{\pi}{t_0} + s_1\right)t - \dfrac{2\pi s_1}{t_0\left(\dfrac{\pi^2}{t_0^2} - s_1^2\right)}\right] \\[4pt] or \\[4pt] s_1\cos s_1 t\left[\dfrac{\cos(\pi - s_1 t_0)}{\pi\big/t_0 - s_1} - \dfrac{\cos(\pi + s_1 t_0)}{\pi\big/t_0 + s_1}\right] - \dfrac{2\pi s_1}{t_0\left(\dfrac{\pi^2}{t_0^2} - s_1^2\right)} \\[4pt] -\sin s_1 t\left[\dfrac{\pi}{\pi - s_1 t_0}\sin\left(\dfrac{\pi}{t_0} - s_1\right)t - \dfrac{\pi}{\pi + s_1 t_0}\sin\left(\dfrac{\pi}{t_0} + s_1\right)t\right], \quad t < t_0 \\[4pt] or \\[4pt] s_1\sin s_1 t\left[\dfrac{\sin(\pi - s_1 t_0)}{\pi\big/t_0 - s_1} - \dfrac{\sin(\pi + s_1 t_0)}{\pi\big/t_0 + s_1}\right], \quad t \geq t_0\end{array}\right\}_{S_1}\sum_{i=1}^{2}\begin{Bmatrix}A_{i_1}\\B_{i_1}\end{Bmatrix}\begin{bmatrix}e_{i_1}\\e_{i_2}\end{bmatrix}$$

$$-\frac{1}{2s_2}\left\{\qquad\qquad\right\}_{S_2}\sum_{i=1}^{2}\begin{Bmatrix}A_{i_2}\\B_{i_2}\end{Bmatrix}\begin{bmatrix}e_{i_1}\\e_{i_2}\end{bmatrix}$$

$$(66')$$

We obtain as in (a)

$$\text{Half sine} \begin{cases} \hat{\gamma}_{j_z} = \dfrac{\hat{Z}_{j_{Max}}}{Z_{M_S}} \\[3em] \hat{\gamma}_{j_\theta} = \dfrac{\hat{\theta}_{j_{Max}}}{\theta_{Y_S}} \end{cases}; \qquad j = 1,2$$

These dynamic response factors are the final results for the Z plane.

2.5   X Axis Analysis

2.5.1   Solution for X axis Transient Response

The equation of motion for the simplified model is given by Eq. (11):

$$\theta_X + \alpha_{23} I_{X_M} \ddot{\theta}_X \;\; = \;\; \alpha_{33} T_X(t) \tag{11'}$$

Assuming $\theta_X = \dot{\theta}_X = 0$ as initial conditions, operating of $(11')$ with the Laplacian operator gives:

$$w + \alpha_{33} I_{X_M} s^2 w = \alpha_{33} h(s) \tag{67}$$

Where
$$
\begin{aligned}
L\{\theta_X(t)\} &= w(s) \\
L\{T_X(t)\} &= h(s)
\end{aligned}
$$

Hence,

$$w = \frac{\alpha_{33} h(s)}{1 + \alpha_{33} I_{X_M} s^2} = \frac{\alpha_{33} h(s)}{\Delta} \tag{68}$$

The characteristic Eq. is $\Delta = 0$

i.e.,
$$1 + \alpha_{33} I_{X_M} s^2 = 0$$

$$f = \alpha_{33} I_{X_M}$$

Let
$$f\left(s^2 + \frac{1}{f}\right) = 0$$

Which leads to the eigenvalue $S_N^2 = \dfrac{1}{f}$

Hence,

$$f \cdot w = \frac{\alpha_{33} h(s)}{\left(s^2 + s_N^2\right)} \tag{69}$$

Taking the inverse transformation, we obtain:

$$
\begin{aligned}
f \cdot \theta_X &= \alpha_{33} L^{-1}\left\{ \frac{h(s)}{s^2 + s_N^2} \right\} \\
&= \frac{\alpha_{33}}{s_N} \int_0^t T_X(\beta) \sin s_N(t - \beta)\, d\beta
\end{aligned}
\tag{70}
$$

using the convolution theorem.

For the three types of functions given on pgs. 13, 14, & 15, the solutions are as follows:

Solutions (a) : T_x applied in square step

$$\theta_X = \alpha_{33}T_{X_{Max}}\begin{cases} 1-\cos s_N t \quad , \quad t < t_0 \\ or \\ \cos s_N(t-t_0) - \cos s_N t \quad , \quad t \geq t_0 \end{cases}$$ (71)

Solution (b) : T_x applied in ramp

$$\theta_X = \frac{\alpha_{33}T_{X_{Max}}}{s_N t_0}\begin{cases} s_N t - \sin s_N t \quad ; \quad t < t_0 \\ or \\ s_N t \cos s_N(t-t_0) - \sin s_N t - s_N(t-t_0)\cos s_N(t-t_0) + \sin s_N(t-t_0); \quad t \geq t_0 \end{cases}$$ (72)

Solution (c) : T_x applied in half sine

$$\theta_X = \frac{-\alpha_{33}T_{X_{Max}}}{2s_N f}\left\{ \begin{array}{c} \sin s_N t\left[\dfrac{\cos\left(\dfrac{\pi}{t_0} - s_N\right)t}{\pi/t_0 - s_N} + \dfrac{\cos\left(\dfrac{\pi}{t_0} + s_N\right)t}{\pi/t_0 + s_N} - \dfrac{2\pi}{t_0\left(\dfrac{\pi^2}{t_0^2} - s_N\right)}\right] \\ + \cos s_N t\left[\dfrac{\sin\left(\dfrac{\pi}{t_0} - s_N\right)t}{\pi/t_0 - s_N} - \dfrac{\sin\left(\dfrac{\pi}{t_0} + s_N\right)t}{\pi/t_0 + s_N}\right]; \quad t < t_0 \\ or \\ \sin s_N t\left[\dfrac{\cos(\pi - t_0 s_N)}{\pi/t_0 - s_N} + \dfrac{\cos(\pi + t_0 s_N)}{\pi/t_0 + s_N} - \dfrac{2\pi}{t_0\left(\dfrac{\pi^2}{t_0^2} - s_N^2\right)}\right] \\ + \cos s_N t\left[\dfrac{\sin(\pi - t_0 s_N)}{\pi/t_0 - s_N} - \dfrac{\sin(\pi + t_0 s_N)}{\pi/t_0 + s_N}\right]; \quad t \geq t_0 \end{array} \right.$$ (73)

2.5.2 Dynamic response factors

Static deflection under the peak transient loading $T_{X_{Max}} = 1$ is obtained from Eq. (11´):

$$\theta_{X_s} = \alpha_{33}$$

The maximum transient response to the forcing function of (a), (b), and (c) will be obtained in a similar manner as for the Y and Z planes. However, the dynamic response factors for these types of inputs, in terms of the ratios of application times to natural period, are well known. Thus, the factors calculated by the use of Eqs. (71), (72), & (73) can be easily checked, providing a check on the terms used elsewhere in the analysis.

Solution (a)

Maximizing Eq. (71), we obtain:

$$0 = \begin{cases} s_N \sin s_N t , & t < t_0 \\ -s_N \sin s_N (t - t_0) + s_N \sin s_N t, & t \geq t_0 \end{cases}$$

(71́)

$$s_N \sin s_N t = 0 \quad for \quad s_N t = m\pi; \quad m = 1,2,....$$

For t < t₀:

$$\therefore \quad t_0 = \frac{m\pi}{s_N}$$

(valid if $t_1 < t_0$)　　(74)

For $t \geq t_0$: $-s_N \sin s_N (t - t_0) + s_N \sin s_N t = 0$

$$- s_N \left[\sin s_N t \cos s_N t_0 - \cos s_N t \sin s_N t_0 \right] + s_N \sin s_N t = 0$$

$$s_N \sin s_N t \left[1 - \cos s_N t_0 \right] + s_N \cos s_N t \sin s_N t_0 = 0$$

$$\tan s_n t = -\frac{\sin s_N t_0}{1 - \cos s_N t_0}$$

Thus, $\quad t_2 = \frac{1}{s_N} \tan^{-1} \left\{ \frac{-\sin s_N t_0}{1 - \cos s_N t_0} \right\}$ 　　(valid if $t_1 \geq t_0$)　　(75)

Substitute into Eq. (71) with $T_{X_{Max}} = 1;$ the maximum response will be obtained. Denoting the maximax (maximum of the maximum) displacement by $\overset{\diamond}{\theta}_{X_{Max}}$, we obtain the dynamic response factor:

Square step $\left\{ \overset{\diamond}{\gamma}_{\theta_X} = \overset{\diamond}{\theta}_{X_{Max}} / \theta_{X_S} \right\}$

(76)

Solution (b)

Maximizing Eq. (72):

$$0 = \begin{cases} s_N - s_N \cos s_N t = 0, & t < t_0 \\ or \\ s_N \cos s_N (t - t_0) - s_N \cos s_N t - s_N^2 t_0 \sin s_n (t - t_0), & t \geq t_0 \end{cases}$$

(72́)

For t < t₀:

$$s_N - s_N \cos s_N t = 0$$
$$s_N (1 - \cos s_N t) = 0$$
$$\cos s_N t = 1$$
$$s_N t = m2\pi; \qquad m = 1,2,...$$
$$\therefore \quad t_1 = \frac{2m\pi}{s_N} \qquad \text{(valid if } t_1 < t_0) \tag{77}$$

For $t \geq t_0$:

$$s_N \cos s_N (t - t_0) - s_N \cos s_N t - s_N^2 t_0 \sin s_N (t - t_0) = 0$$

Expanding:

$$0 = s_N \left[\cos s_N t \cos s_N t_0 + \sin s_N t \sin s_N t_0 \right] - s_N \cos s_N t - s_N^2 t_0 \left[\sin s_N t \cos s_N t_0 - \cos s_N t \sin s_N t_0 \right]$$

$$\sin s_N t \left[s_N \sin s_N t_0 - s_N^2 t_0 \cos s_N t \right] = \cos s_N t \left[s_N - s_N \cos s_N t_0 - s_N^2 t_0 \sin s_N t_0 \right]$$

$$\tan s_N t = \frac{s_N - s_N \cos s_N t_0 - s_N^2 t_0 \sin s_N t_0}{s_N \sin s_N t_0 - s_N^2 t_0 \cos s_N t_0}$$

$$\therefore \quad t_2 = \frac{1}{s_N} \tan^{-1} \left[\frac{s_N - s_N \cos s_N t_0 - s_N^2 t_0 \sin s_N t_0}{s_N \sin s_N t_0 - s_N^2 t_0 \cos s_N t_0} \right] \qquad \text{(valid if } t_1 \geq t_0) \tag{78}$$

Substitute into Eq. (72), we obtain $\overset{\Delta}{\theta}_{X_{Max}}$, and

$$\text{Ramp} \left\{ \overset{\Delta}{\gamma}_{\theta_X} = \overset{\Delta}{\theta}_{X_{Max}} / \theta_{X_S} \right. \tag{79}$$

Solution (c):

Equation (73) can be simplified as follows:

For $t < t_0$: $\qquad\qquad \theta_X = -\frac{\alpha_{33} T_{X_{Max}}}{2 s_N \frac{1}{s_N^2}} \left\{ \right\}$

43

$$
\{\ \} = \sin s_N t \left\{
\begin{array}{l}
\dfrac{t_0}{\pi - s_N t_0}\left[\cos\dfrac{\pi t}{t_0}\cos s_N t + \sin\dfrac{\pi t}{t_0}\sin s_N t\right] \\[2mm]
+\dfrac{t_0}{\pi + s_N t_0}\left[\cos\dfrac{\pi t}{t_0}\cos s_N t - \sin\dfrac{\pi t}{t_0}\sin s_N t\right] - \dfrac{2\pi}{\pi^2 - s_N^2 t_0}
\end{array}
\right\}
$$

$$
+\cos s_N t \left\{
\begin{array}{l}
\dfrac{t_0}{\pi - s_N t_0}\left[\sin\dfrac{\pi t}{t_0}\cos s_N t - \cos\dfrac{\pi t}{t_0}\sin s_N t\right] \\[2mm]
-\dfrac{t_0}{\pi + s_N t_0}\left[\sin\dfrac{\pi t}{t_0}\cos s_N t + \cos\dfrac{\pi t}{t_0}\sin s_N t\right]
\end{array}
\right\}
$$

$$
= \sin s_N t\left\{\cos\dfrac{\pi t}{t_0}\cos s_N t\underbrace{\left(\dfrac{t_0}{\pi - s_N t}+\dfrac{t_0}{\pi + s_N t}\right)}_{C} + \sin\dfrac{\pi t}{t_0}\sin s_N t\underbrace{\left(\dfrac{t_0}{\pi - s_N t}-\dfrac{t_0}{\pi + s_N t}\right)}_{D}\right.
$$

$$
\left.-\dfrac{2\pi}{\pi^2 - s_N^2 t_0}\right\} + \cos s_N t\left\{\sin\dfrac{\pi t}{t_0}\cos s_N t\ (\overset{D}{\ })-\cos\dfrac{\pi t}{t_0}\sin s_N t\ (\overset{C}{\ })\right\}
$$

$$
= \sin s_N t\cos s_N t\cos\dfrac{\pi t}{t_0}\cdot C + \sin^2 s_N t\sin\dfrac{\pi t}{t_0}\cdot D - \dfrac{2\pi t_0}{\pi^2 - s_N^2 t_0^2}\sin s_N t
$$

$$
+\cos^2 s_N t\sin\dfrac{\pi t}{t_0}\cdot D - \sin s_N t\cos s_N t\cos\dfrac{\pi t}{t_0}\cdot C
$$

or,
$$
\theta_X = -\dfrac{s_N\alpha_{33}T_{X_{Max}}}{2}\left\{D\cdot\sin\dfrac{\pi t}{t_0}-\dfrac{2\pi t_0}{\pi^2 - s_N^2 t_0^2}\sin s_N t\right\}
$$

Now,
$$
D = \dfrac{t_0(\pi + s_N t_0)-t_0(\pi - s_N t_0)}{\pi^2 - s_N^2 t_0^2}=\dfrac{2 s_N t_0^2}{\pi^2 - s_N^2 t_0^2}
$$

$$
\therefore\quad \theta_X = \alpha_{33}T_{X_{Max}}\dfrac{s_N t_0}{\pi^2 - s_N^2 t_0^2}\left\{\pi\cdot\sin s_N t - s_N t_0\sin_N\dfrac{\pi t}{t_0}\right\},\quad t < t_0 \tag{73$'$}
$$

Maximizing:

$$
\dfrac{d\theta_X}{dt}=0=\pi s_N\cos s_N t_1 - \pi s_N\cos\dfrac{\pi t_1}{t_0}
$$

Hence, for $t_1 < t_0$:

$$
\cos s_N t_1 = \cos\dfrac{\pi t_1}{t_0}
$$

Thus,
$$
\dfrac{\pi t_1}{t_0}=2m\pi \pm s_N t_1 \quad ; m=1,2,\ldots
$$

Or,
$$
t_1 = \dfrac{2m\pi t_0}{\pi \mp s_N t_0}\qquad\qquad ; m=1,2,\ldots \tag{80}
$$

$$
\text{(valid for } t_1 < t_0)
$$

Also, for $t \geq t_0$, from Eq. (73):

$$\left\{ \begin{array}{l} \\ \end{array} \right\} = \sin s_N t \left\{ \frac{t_0}{\pi - s_N t_0} \left[\cos \pi \cos s_N t_0 + \sin \pi \sin s_N t_0 \right] \right.$$

$$+ \frac{t_0}{\pi + s_N t_0} \left[\cos \pi \cos s_N t_0 - \sin \pi \sin s_N t_0 \right] - \left. \frac{2\pi t_0}{\pi^2 - s_N^2 t_0^2} \right\}$$

$$+ \cos s_N t \left\{ \frac{t_0}{\pi - s_N t_0} \left[\sin \pi \cos s_N t_0 - \cos \pi \sin s_N t_0 \right] - \frac{t_0}{\pi + s_N t_0} \left[\sin \pi \cos s_N t_0 + \cos \pi \sin s_N t_0 \right] \right\} \text{Thus,}$$

$$= \sin s_N t \left[\cos \pi \cos s_N t_0 \left(\overset{C}{} \right) - \frac{2\pi t_0}{\pi^2 - s_N^2 t_0^2} \right] + \cos s_N t \left[- \cos \pi \sin s_N t_0 \left(\overset{C}{} \right) \right]$$

$$= - \sin s_N t \cos s_N t \cdot C - \sin s_N t \frac{2\pi t_0}{\pi^2 - s_N^2 t_0^2} + \cos s_N t \sin s_N t_0 \cdot C$$

$$\theta_X = - \frac{s_N \alpha_{33} T_{X_{Mac}}}{2} \left\{ C \left[\cos s_N t \sin s_N t_0 - \sin s_N t \cos s_N t_0 \right] - \frac{2\pi t_0 s_N}{\pi^2 - s_N^2 t_0^2} \right\} \tag{73''}$$

Now, $\qquad C = \dfrac{t_0(\pi + s_N t_0) + t_0(\pi - s_N t_0)}{\pi^2 - s_N^2 t_0^2} = \dfrac{2 t_0 \pi}{\pi^2 - s_N^2 t_0^2}$

$$\therefore \theta_X + \alpha_{33} T_{X_{Max}} \cdot \frac{\pi s_N t_0}{\pi^2 - s_N^2 t_0^2} \left\{ \sin s_N t (1 + \cos s_N t_0) - \cos s_N t \sin s_N t_0 \right\}, \quad t \geq t_0$$

Maximizing:

$$\frac{d\theta_X}{dt} = 0 = s_N (1 + \cos s_N t_0) \cos s_N t_2 + s_N \sin s_N t_0 \sin s_N t_2$$

Hence, for $t_2 \geq t_0$:

$$t_2 = \frac{1}{s_N} \tan^{-1} \left\{ \frac{-(1 + \cos s_N t_0)}{\sin s_N t_0} \right\} \quad , \quad (\text{valid for } t_2 \geq t_0) \tag{81}$$

Substitute into Eqs. (73') & (73'') with $T_{X_{Max}} = 1$, we obtain $\hat{\theta}_{X_{Max}}$, then,

Half sine $\left\{ \hat{\gamma}_{\theta_X} = \hat{\theta}_{X_{Max}} / \hat{\theta}_{X_S} \right.$ \hfill (82)

2.6 Mass and Inertia Properties

Y Plane:

$$M = \frac{1}{4}M_{Arm} + M_{Gondola} + M_{Gimbal} + M_{Forks}$$

$$I_{Z_M} = I_{Z_{Forks}} + I_{Z_{Gondola}} + I_{Z_{Gimbal}}$$

$$I_{Z_R} = I_{Z_{Tot}} - I_{Z_M} - MR^2$$

Z Plane:

$$M = M_{Y\,Plane}$$

$$I_{Y_M} = I_{Y_{Gondola}} + I_{Y_{Gimbal}}$$

X Plane:

$$I_{X_M} = I_{X_{Force}} + I_{X_{Gondola}} + I_{X_{Gimbal}} + \frac{1}{4}I_{X_{Arm}}$$

Mass and Inertia Properties of Centrifuge

Body	M $lb.\sec^2 in^{-1}$	I_X $lb.in.\sec^2$	I_Y $lb.in.\sec^2$	I_Z $lb.in.\sec^2$
Arm	55.4	210.0E+3	-	-
Forks	6.2	73.0E+3	-	72.8E+3
Gondola	18.5	26.8E+3	41.2E+3	62.9E+3
Gimbal	48.9	207.7E+3	189.8E+3	388.1E+3
Total	-	-	-	42.0E+6

TABLE I

2.7   Numerical Calculations: X–axis

Parameters: $\alpha_{33} = \dfrac{46.31 \times 10^{-6}}{108 \times 216}$ (Ref. Appendix 2, Eq. (8))

$I_{X_M} = 360.0 \times 10^3$ (Ref. Page 41)

From page 35:

$$s_N = \sqrt{\frac{1}{f}} = \frac{1}{\sqrt{\alpha_{33} I_{X_M}}} = \sqrt{1399} = 37.4 \quad Rad/sec$$

Hence, the natural frequency for this simplified model is:

$$f_m = \frac{s_N}{2\pi} = \underline{5.95} \text{ cps}$$

And the natural period is :

$$T = \frac{1}{f_m} = \underline{0.168} \text{ sec.}$$

Solution (a):

From Eq. (74): $t_1 = \dfrac{m\pi}{s_N} = .084 \ ; \ m = 1,2,3,...$ $\left(for \ \ t_1 < t_0\right)$

From Eq. (75): $t_2 = \dfrac{1}{s_N} \tan^{-1}\left\{\dfrac{-\sin s_N t_0}{1 - \cos s_N t_0}\right\}$ $,\left(for \ \ t_1 \geq t_0\right)$

Tables I and II, pages 42 and 43, give a tabular computation of the times at which maxima occur and the dynamic response factors, obtained from Eqs. (71) & (76), for various ratios t_0/T, where:

$$\gamma_{\theta_X}^{\diamond} = \begin{cases} 1 - \cos s_N t_1 & ; t_1 < t_0 \\ \cos s_N(t_2 - t_0) - \cos s_N t_2 & ; t_2 \geq t_0 \end{cases}$$

to/t	to	sn*to	sin (sn to)	cos(sn to)	1-cos(sn to)	minus D/F	arctan(G)	Pi-H	TIME
0.02	0.00336	0.12566	0.12533	0.99211	0.00789	-15.89451	1.50796	1.63363	0.04368
0.04	0.00672	0.25133	0.24869	0.96858	0.03142	-7.91580	1.44513	1.69646	0.04536
0.06	0.01008	0.37699	0.36813	0.92978	0.07022	-5.24217	1.38230	1.75929	0.04704
0.08	0.01344	0.50266	0.48175	0.87631	0.12369	-3.89473	1.31947	1.82212	0.04872
0.10	0.01680	0.62832	0.58779	0.80902	0.19098	-3.07768	1.25664	1.88496	0.05040
0.12	0.02016	0.75398	0.68455	0.72897	0.27103	-2.52571	1.19380	1.94779	0.05208
0.14	0.02352	0.87965	0.77051	0.63742	0.36258	-2.12510	1.13097	2.01062	0.05376
0.16	0.02688	1.00531	0.84433	0.53582	0.46418	-1.81899	1.06814	2.07345	0.05544
0.18	0.03024	1.13098	0.90483	0.42578	0.57422	-1.57574	1.00531	2.13628	0.05712
0.20	0.03360	1.25664	0.95106	0.30901	0.69099	-1.37638	0.94248	2.19912	0.05880
0.22	0.03696	1.38230	0.98229	0.18738	0.81262	-1.20879	0.87964	2.26195	0.06048
0.24	0.04032	1.50797	0.99803	0.06279	0.93721	-1.06489	0.81681	2.32478	0.06216
0.26	0.04368	1.63363	0.99803	-0.06279	1.06279	-0.93906	0.75398	2.38761	0.06384
0.28	0.04704	1.75930	0.98229	-0.18739	1.18739	-0.82727	0.69115	2.45044	0.06552
0.30	0.05040	1.88496	0.95106	-0.30902	1.30902	-0.72654	0.62832	2.51328	0.06721
0.32	0.05376	2.01062	0.90483	-0.42578	1.42578	-0.63462	0.56548	2.57611	0.06889
0.34	0.05712	2.13629	0.84433	-0.53583	1.53583	-0.54975	0.50265	2.63894	0.07057
0.36	0.06048	2.26195	0.77051	-0.63743	1.63743	-0.47056	0.43982	2.70177	0.07225
0.38	0.06384	2.38762	0.68454	-0.72897	1.72897	-0.39592	0.37699	2.76460	0.07393
0.40	0.06720	2.51328	0.58778	-0.80902	1.80902	-0.32492	0.31416	2.82744	0.07561
0.42	0.07056	2.63894	0.48175	-0.87631	1.87631	-0.25675	0.25132	2.89027	0.07729
0.44	0.07392	2.76461	0.36812	-0.92978	1.92978	-0.19076	0.18849	2.95310	0.07897
0.46	0.07728	2.89027	0.24868	-0.96858	1.96858	-0.12633	0.12566	3.01593	0.08065
0.48	0.08064	3.01594	0.12533	-0.99212	1.99212	-0.06291	0.06283	3.07876	0.08233
0.50	0.08400	3.14160	-0.00001	-1.00000	2.00000	0.00000	0.00000	3.14160	0.08401

TIMES AT WHICH MAXIMUM RESPONSE OCCURS- SQUARE STEP

TABLE II

to/t	t_2	$37.4*t_2$		cos(D)	cos(C)	γ_{ex}
0.02	0.04368	1.63363	1.50797	0.06279	-0.06279	1.06279
0.04	0.04536	1.69646	1.44514	0.12533	-0.12534	1.12534
0.06	0.04704	1.75930	1.38230	0.18738	-0.18739	1.18739
0.08	0.04872	1.82213	1.31947	0.24869	-0.24869	1.24869
0.10	0.05040	1.88496	1.25664	0.30901	-0.30902	1.30902
0.12	0.05208	1.94779	1.19381	0.36812	-0.36813	1.36813
0.14	0.05376	2.01062	1.13098	0.42578	-0.42578	1.42578
0.16	0.05544	2.07346	1.06814	0.48175	-0.48176	1.48176
0.18	0.05712	2.13629	1.00531	0.53582	-0.53583	1.53583
0.20	0.05880	2.19912	0.94248	0.58778	-0.58779	1.58779
0.22	0.06048	2.26195	0.87965	0.63742	-0.63743	1.63743
0.24	0.06216	2.32478	0.81682	0.68455	-0.68455	1.68455
0.26	0.06384	2.38762	0.75398	0.72897	-0.72897	1.72897
0.28	0.06552	2.45045	0.69115	0.77051	-0.77052	1.77052
0.30	0.06720	2.51328	0.62832	0.80902	-0.80902	1.80902
0.32	0.06888	2.57611	0.56549	0.84433	-0.84433	1.84433
0.34	0.07056	2.63894	0.50266	0.87631	-0.87631	1.87631
0.36	0.07224	2.70178	0.43982	0.90483	-0.90483	1.90483
0.38	0.07392	2.76461	0.37699	0.92978	-0.92978	1.92978
0.40	0.07560	2.82744	0.31416	0.95106	-0.95106	1.95106
0.42	0.07728	2.89027	0.25133	0.96858	-0.96858	1.96858
0.44	0.07896	2.95310	0.18850	0.98229	-0.98229	1.98229
0.46	0.08064	3.01594	0.12566	0.99211	-0.99212	1.99212
0.48	0.08232	3.07877	0.06283	0.99803	-0.99803	1.99803
0.50	0.08400	3.14160	0.00000	1.00000	-1.00000	2.00000

DYNAMIC RESPONSE FACTORS- SQUARE STEP

TABLE III

Solution (b):

From Eq. (77): $t_1 = \dfrac{2m\pi}{s_N} = .168m$; $m = 1,2,3,....,$ (valid if $t_1 < t_0$)

From Eq. (78): $t_2 = \dfrac{1}{37.4}\tan^{-1}\left\{\dfrac{37.4 - 37.4\cos 37.4t_0 - 1398.8t_0 \sin 37.4t_0}{37.4\sin 37.4 - 1398.8t_0 \cos 37.4t_0}\right\}$

(valid if $t_1 \geq t_0$)

Table III on page 46 calculates the values of t_1 & t_2 at which maxima occur for various ratios t_0/T.

Table IV, page 47, gives the dynamic response factors for these ratios from Eqs. (72) & (79), using the values $t_{1,2}$ from Table III.

Solution (c):

From Eq. (80): $t_1 = \dfrac{2m\pi t_0}{\pi + 37.4t_0}$; m=1, 2, 3, ..., (valid if $t_1 < t_0$)

Note: the roots t_1 corresponding to the use of the negative sign in Eq. (80) result in minimum values. Hence, they may be ignored.

From Eq. (81): $t_2 = \dfrac{1}{37.4}\tan^{-1}\left\{\dfrac{-(1 + \cos 37.4t_0)}{\sin 37.4t_0}\right\}$,   (valid if $t_2 \geq t_1$)

Table V on page 48 gives the dynamic values of t_1 and t_2 at which maxima occur for the ratio t_0/T.

Table VI, pages 49 & 50, gives the dynamic response factors for these ratios, calculated from Eqs. (73)́, (73)″, & (82), using the values $t_{1,2}$ from Table V.

		phi	1	2	3	4	5
0.1	0.0168	0.6283	0.8090	0.5878	30.2572	21.9832	23.4998
0.2	0.0336	1.2566	0.3090	0.9511	11.5571	35.5695	46.9997
0.3	0.0504	1.8850	-0.3090	0.9511	-11.5574	35.5695	70.4995
0.4	0.0672	2.5133	-0.8090	0.5878	-30.2574	21.9830	93.9994
0.5	0.0840	3.1416	-1.0000	0.0000	-37.4000	-0.0003	117.4992
0.6	0.1008	3.7699	-0.8090	-0.5878	-30.2570	-21.9834	140.9990
0.7	0.1176	4.3982	-0.3090	-0.9511	-11.5569	-35.5696	164.4989
0.8	0.1344	5.0266	0.3090	-0.9511	11.5577	-35.5694	187.9987
0.9	0.1512	5.6549	0.8090	-0.5878	30.2575	-21.9828	211.4986
1	0.1680	6.2832	1.0000	0.0000	37.4000	0.0005	234.9984
1.1	0.1848	6.9115	0.8090	0.5878	30.2569	21.9837	258.4982
1.2	0.2016	7.5398	0.3090	0.9511	11.5566	35.5697	281.9981
1.3	0.2184	8.1682	-0.3090	0.9511	-11.5579	35.5693	305.4979
1.4	0.2352	8.7965	-0.8090	0.5878	-30.2577	21.9825	328.9978
1.5	0.2520	9.4248	-1.0000	0.0000	-37.4000	-0.0008	352.4976

to/t							t2
	6	7	8	9	10	PSI	PSI/37.4
0.1	13.8129	19.0117	-6.6701	2.9715	-2.2447	1.9899	0.0532
0.2	44.6994	14.5236	-18.8565	21.0460	-0.8960	2.4110	0.0645
0.3	67.0489	-21.7858	-18.0915	57.3553	-0.3154	2.8360	0.0758
0.4	55.2510	-76.0474	12.4064	98.0304	0.1266	3.2675	0.0874
0.5	-0.0009	-117.4992	74.8009	117.4989	0.6366	3.7085	0.0992
0.6	-82.8782	-114.0699	150.5352	92.0865	1.6347	4.1634	0.1113
0.7	-156.4483	-50.8313	205.4051	15.2617	13.4589	4.6382	0.1240
0.8	-178.7967	58.0969	204.6391	-93.6663	-2.1848	2.0001	0.0535
0.9	-124.3135	171.1076	131.4559	-193.0903	-0.6808	5.6855	0.1520
1	0.0035	234.9984	-0.0035	-234.9979	0.0000	6.2832	0.1680
1.1	151.9448	209.1270	-144.8017	-187.1434	0.7737	6.9417	0.1856
1.2	268.1976	87.1375	-242.3543	-51.5678	4.6997	7.6443	0.2044
1.3	290.5440	-94.4096	-241.5861	129.9789	-1.8587	8.3476	0.2232
1.4	193.3746	-266.1688	-125.7169	288.1513	-0.4363	9.0134	0.2410
1.5	-0.0078	-352.4976	74.8078	352.4968	0.2122	9.6339	0.2576

TIMES AT WHICH MAXIMUM RESPONSE OCCURS- RAMP INPUT

TABLE IV

to/t	t2		phi	1-2	cos 3	sin 3	1 x 4
	PSI/37.4	1	2	3	4	5	6
0.1	0.0532	1.9899	0.6283	1.3616	0.2077	0.9782	0.4133
0.2	0.0645	2.4110	1.2566	1.1544	0.4045	0.9145	0.9752
0.3	0.0758	2.8360	1.8850	0.9511	0.5808	0.8140	1.6472
0.4	0.0874	3.2675	2.5133	0.7542	0.7288	0.6847	2.3814
0.5	0.0992	3.7085	3.1416	0.5669	0.8436	0.5370	3.1284
0.6	0.1113	4.1634	3.7699	0.3935	0.9236	0.3834	3.8452
0.7	0.1240	4.6382	4.3982	0.2400	0.9713	0.2377	4.5053
0.8	0.0535	2.0001	5.0266	-3.0265	-0.9934	-0.1148	-1.9868
0.9	0.1520	5.6855	5.6549	0.0306	0.9995	0.0306	5.6828
1	0.1680	6.2832	6.2832	0.0000	1.0000	0.0000	6.2832
1.1	0.1856	6.9417	6.9115	0.0302	0.9995	0.0302	6.9386
1.2	0.2044	7.6443	7.5398	0.1045	0.9945	0.1043	7.6026
1.3	0.2232	8.3476	8.1682	0.1794	0.9839	0.1785	8.2136
1.4	0.2410	9.0134	8.7965	0.2169	0.9766	0.2152	8.8022
1.5	0.2576	9.6339	9.4248	0.2091	0.9782	0.2076	9.4240

to/t				$\gamma_{\theta x}$
	7	8	9	9/2
0.1	0.9135	0.2828	0.1952	0.3107
0.2	0.6673	0.4669	0.7555	0.6012
0.3	0.3008	0.5524	1.6080	0.8531
0.4	-0.1256	0.5497	2.6420	1.0512
0.5	-0.5370	0.4782	3.7242	1.1854
0.6	-0.8531	0.3634	4.7183	1.2516
0.7	-0.9973	0.2331	5.5071	1.2521
0.8	0.9093	3.0065	-6.0174	-1.1971
0.9	-0.5628	0.0306	6.2456	1.1045
1	0.0000	0.0000	6.2832	1.0000
1.1	0.6120	0.0302	6.3266	0.9154
1.2	0.9781	0.1039	6.6249	0.8787
1.3	0.8806	0.1766	7.3349	0.8980
1.4	0.3999	0.2118	8.4057	0.9556
1.5	-0.2076	0.2046	9.6347	1.0223

DYNAMIC RESPONSE FACTORS- RAMP INPUT

TABLE V

to/t	to		sin 1	cos 1	1.+3	-4/2	atan 5
		1	2	3	4	5	6
0.1	0.0168	0.6283	0.5878	0.8090	1.8090	-3.0777	1.8850
0.2	0.0336	1.2566	0.9511	0.3090	1.3090	-1.3764	2.1991
0.3	0.0504	1.8850	0.9511	-0.3090	0.6910	-0.7265	2.5133
0.4	0.0672	2.5133	0.5878	-0.8090	0.1910	-0.3249	2.8274
0.5	0.0840	3.1416	0.0000	-1.0000	0.0000	0.0000	3.1416
0.6	0.1008	3.7699	-0.5878	-0.8090	0.1910	0.3249	3.4558
0.7	0.1176	4.3982	-0.9511	-0.3090	0.6910	0.7266	3.7699
0.8	0.1344	5.0266	-0.9511	0.3090	1.3090	1.3764	4.0841
0.9	0.1512	5.6549	-0.5878	0.8090	1.8090	3.0778	4.3982
1	0.1680	6.2832	0.0000	1.0000	2.0000	9999999.0000	4.7124
1.1	0.1848	6.9115	0.5878	0.8090	1.8090	-3.0776	5.0266
1.2	0.2016	7.5398	0.9511	0.3090	1.3090	-1.3764	5.3407
1.3	0.2184	8.1682	0.9511	-0.3090	0.6910	-0.7265	5.6549
1.4	0.2352	8.7965	0.5878	-0.8090	0.1910	-0.3249	5.9691
1.5	0.2520	9.4248	0.0000	-1.0000	0.0000	0.0000	6.2832

to/t	t_2 6/37.4	$\pi+1$	2π x to	t_1
	7	8	9	10
0.1	0.0504	3.7699	0.1056	0.0280
0.2	0.0588	4.3982	0.2111	0.0480
0.3	0.0672	5.0266	0.3167	0.0630
0.4	0.0756	5.6549	0.4222	0.0747
0.5	0.0840	6.2832	0.5278	0.0840
0.6	0.0924	6.9115	0.6333	0.0916
0.7	0.1008	7.5398	0.7389	0.0980
0.8	0.1092	8.1682	0.8445	0.1034
0.9	0.1176	8.7965	0.9500	0.1080
1	0.1260	9.4248	1.0556	0.1120
1.1	0.1344	10.0531	1.1611	0.1155
1.2	0.1428	10.6814	1.2667	0.1186
1.3	0.1512	11.3098	1.3723	0.1213
1.4	0.1596	11.9381	1.4778	0.1238
1.5	0.1680	12.5664	1.5834	0.1260

TIMES FOR MAXIMA- HALF SINE

TABLE VI

to/t	to		1^2	PI^2-2	1/3	t1	37.4*5
		1	2	3	4	5	6
0.1	0.0168	0.6283	0.3948	9.4749	0.0663	0.0280	-
0.2	0.0336	1.2566	1.5791	8.2905	0.1516	0.0480	-
0.3	0.0504	1.8850	3.5531	6.3166	0.2984	0.0630	-
0.4	0.0672	2.5133	6.3166	3.5531	0.7074	0.0747	-
0.5	0.0840	3.1416	9.8697	0.0000	99999.0000	0.0840	3.1416
0.6	0.1008	3.7699	14.2123	-4.3426	-0.8681	0.0916	3.4272
0.7	0.1176	4.3982	19.3445	-9.4749	-0.4642	0.0980	3.6652
0.8	0.1344	5.0266	25.2663	-15.3967	-0.3265	0.1034	3.8666
0.9	0.1512	5.6549	31.9777	-22.1080	-0.2558	0.1080	4.0392
1	0.1680	6.2832	39.4786	-29.6090	-0.2122	0.1120	4.1888
1.1	0.1848	6.9115	47.7691	-37.8995	-0.1824	0.1155	4.3197
1.2	0.2016	7.5398	56.8492	-46.9795	-0.1605	0.1186	4.4352
1.3	0.2184	8.1682	66.7188	-56.8492	-0.1437	0.1213	4.5379
1.4	0.2352	8.7965	77.3781	-67.5084	-0.1303	0.1238	4.6297
1.5	0.2520	9.4248	88.8269	-78.9572	-0.1194	0.1260	4.7124

to/t	SIN 6	$\pi \equiv 7$	π x 5/to	SIN 9	1 x 10	8-11	$\gamma_{\theta x}$
	7	8	9	10	11	12	1*12/3
0.1	-	-	-	-	-	-	-
0.2	-	-	-	-	-	-	-
0.3	-	-	-	-	-	-	-
0.4	-	-	-	-	-	-	-
0.5	0.0000	0.0000	3.1416	0.0000	0.0000	0.0000	1.5708
0.6	-0.2817	-0.8851	2.8560	0.2817	1.0621	-1.9472	1.6904
0.7	-0.5000	-1.5708	2.6180	0.5000	2.1991	-3.7699	1.7500
0.8	-0.6631	-2.0833	2.4166	0.6631	3.3332	-5.4165	1.7683
0.9	-0.7818	-2.4562	2.2440	0.7818	4.4211	-6.8774	1.7591
1	-0.8660	-2.7207	2.0944	0.8660	5.4414	-8.1621	1.7321
1.1	-0.9239	-2.9025	1.9635	0.9239	6.3854	-9.2879	1.6938
1.2	-0.9618	-3.0217	1.8480	0.9618	7.2520	-10.2737	1.6488
1.3	-0.9848	-3.0939	1.7453	0.9848	8.0441	-11.1379	1.6003
1.4	-0.9966	-3.1309	1.6535	0.9966	8.7664	-11.8973	1.5502
1.5	-1.0000	-3.1416	1.5708	1.0000	9.4248	-12.5664	1.5000

DYNAMIC RESPONSE FACTORS- HALF SINE

TABLE VII

to/t	to	pi x 1	13/3	sin 1	cos 1	1+16	t₂ Table V	37.4*18
		13	14	15	16	17	18	19
0.1	0.0168	1.9739	0.2083	0.5878	0.8090	1.8090	0.0504	1.8850
0.2	0.0336	3.9479	0.4762	0.9511	0.3090	1.3090	0.0588	2.1991
0.3	0.0504	5.9218	0.9375	0.9511	-0.3090	0.6910	0.0672	2.5133
0.4	0.0672	7.8957	2.2222	0.5878	-0.8090	0.1910	0.0756	2.8274
0.5	0.0840	9.8697	0.0000	0.0000	-1.0000	0.0000	0.0840	3.1416
0.6	0.1008	-	-	-	-	-	-	
0.7	0.1176	-	-	-	-	-	-	
0.8	0.1344	-	-	-	-	-	-	
0.9	0.1512	-	-	-	-	-	-	
1	0.1680	-	-	-	-	-	-	
1.1	0.1848	-	-	-	-	-	-	
1.2	0.2016	-	-	-	-	-	-	
1.3	0.2184	-	-	-	-	-	-	
1.4	0.2352	-	-	-	-	-	-	
1.5	0.2520	-	-	-	-	-	-	

to/t	SIN 19	COS 19	17 X 20	15 X 21	22 - 23	$\gamma_{\theta x}$
	20	21	22	23	24	14 X 24
0.1	0.9511	-0.3090	1.7205	-0.1816	1.9021	0.3963
0.2	0.8090	-0.5878	1.0590	-0.5590	1.6180	0.7705
0.3	0.5878	-0.8090	0.4061	-0.7694	1.1756	1.1021
0.4	0.3090	-0.9511	0.0590	-0.5590	0.6180	1.3734
0.5	0.0000	-1.0000	0.0000	0.0000	0.0000	1.5714
0.6	-	-	-	-	-	-
0.7	-	-	-	-	-	-
0.8	-	-	-	-	-	-
0.9	-	-	-	-	-	-
1	-	-	-	-	-	-
1.1	-	-	-	-	-	-
1.2	-	-	-	-	-	-
1.3	-	-	-	-	-	-
1.4	-	-	-	-	-	-
1.5	-	-	-	-	-	-

DYNAMIC RESPONSE FACTORS- HALF SINE

TABLE VII (continued)

2.8 Numerical Calculations – Z Axis

$$\left. \begin{array}{l} \alpha_{22} = +39.42 \times 10^{-6} \\ \alpha_{24} = -.135 \times 10^{-6} \quad = \alpha_{42} \\ \alpha_{44} = +.0175 \times 10^{-6} \end{array} \right\}$$

Parameters (Reference Eq. (8), Appendix 2)

$$\left. \begin{array}{l} I_{Y_M} = 231.0 \times 10^3 \\ M = 87.4 \end{array} \right\}$$ (Reference Page 42)

From Page 29:

$$\alpha_{11} = \alpha_{44} I_{Y_M} = .0175 \times 10^{-6} \times 231.0 \times 10^3 = 4.042 \times 10^{-3}$$

$$\alpha_{12} = -\alpha_{24} I_{Y_M} = .135 \times 10^{-6} \times 231.0 \times 10^3 = 31.185 \times 10^{-3}$$

$$\alpha_{21} = -\alpha_{42} M = .135 \times 10^{-6} \times 87.4 = 11.799 \times 10^{-6}$$

$$\alpha_{22} = \alpha_{22} M = 39.42 \times 10^{-6} \times 87.4 = 3445.308 \times 10^{-6}$$

$$d_1 = a_{11} a_{22} - a_{12} a_{21} = 13.558 \times 10^{-6}$$

$$d_2 = a_{11} + a_{22} = 7.487 \times 10^{-3}$$

$$A = \frac{d_2}{d_1} = 552.220$$

$$B = \frac{1}{d_1} = 7.376 \times 10^4$$

From Eq. (56):

$$s^2 = -\frac{A}{2} \pm \sqrt{\left(\frac{A}{2}\right)^2 - B}$$

$$= -\frac{552.220}{2} \pm \sqrt{(276.110)^2 - 7.376 \times 10^4}$$

$$= -276.110 \pm \sqrt{.275 \times 10^4} = -276.110 \pm 52.450$$

As a result:

$$s_1^2 = 223.66, \quad s_1 = 14.956, \quad f_1 = \frac{s_1}{2\pi} = 2.38$$

$$s_2^2 = 328.56, \quad s_2 = 18.126, \quad f_2 = \frac{s_2}{2\pi} = 2.88$$

Also,
$$s_2^2 - s_1^2 = 104.90$$
$$d_1 s_1^2 = 3.032 \times 10^{-3}$$
$$d_2 s_2^2 = 4.455 \times 10^{-3}$$

Evaluating the coefficients of Eq. (58):

$$A_{11} = \frac{1 - 4.042 \times 10^{-3} \times 223.66}{104.90} = 9.148 \times 10^{-4}$$

$$A_{12} = -\frac{1 - 4.042 \times 10^{-3} \times 328.56}{104.90} = 31.272 \times 10^{-4}$$

$$A_{21} = -\frac{31.185 \times 10^{-3} \times 223.66}{104.90} = -664.903 \times 10^{-4}$$

$$A_{22} = \frac{31.185 \times 10^{-3} \times 328.56}{104.90} = 976.753 \times 10^{-4}$$

$$B_{11} = -\frac{11.799 \times 10^{-6} \times 223.66}{104.90} = -25.157 \times 10^{-6}$$

$$B_{12} = \frac{11.799 \times 10^{-6} \times 328.56}{104.90} = 36.956 \times 10^{-6}$$

$$B_{21} = \frac{1 - 3.445 \times 10^{-3} \times 223.66}{104.90} = 21.877 \times 10^{-4}$$

$$B_{22} = -\frac{1 - 3.445 \times 10^{-3} \times 328.56}{104.90} = 12.573 \times 10^{-4}$$

From Eq. (59):

$$e_{11} = 39.42 \times 10^{-6} \qquad e_{12} = -.135 \times 10^{-6}$$
$$e_{21} = -.135 \times 10^{-6} \qquad e_{22} = .0175 \times 10^{-6}$$

Dynamic Response:

The procedure recommended in pages 32 through 34 is not advantageous when t_0 is a variable parameter. In this case it is better to obtain the maximax response directly from Eqs. (64) through (66). For each distinct value assigned to t_0, the functions in the double brackets may be evaluated versus time. We note that if time is scaled correctly, the two functions $f_1(t) = 1 - \cos s_1 t$ and $f_2(t) = 1 - \cos s_2 t$ appear as identical functions; i.e., letting $t^1 = \frac{s_2}{s_1} t$, also $t_{0_K}{}^1 = \frac{s_2}{s_1} t_{0_K}$ to get identical functions.

We have

$$f_2(t) = f_2\left(\frac{s_1 t^1}{s_2}\right) = 1 - \cos s_1 t^1 \equiv f_1(t^1) = f_1\left(\frac{s_2}{s_1}t\right)$$

Hence, the response may be obtained from $f_1(t)$ by the time scaling and multiplication by the appropriate constants. For convenience in notation, let:

$$\Gamma_{11} = \frac{A_{11}e_{11} + A_{21}e_{21}}{d_1 s_1^2} = \frac{[9.148 \times 39.42 + 664.903 \times .135] \times 10^{-10}}{3.032 \times 10^{-3}} = 148.540 \times 10^{-7}$$

$$\Gamma_{12} = \frac{A_{12}e_{12} + A_{22}e_{21}}{d_1 s_2^2} = \frac{[31.272 \times 39.42 - 976.753 \times .135] \times 10^{-10}}{4.455 \times 10^{-3}} = 247.111 \times 10^{-7}$$

$$\Gamma_{21} = \frac{A_{11}e_{12} + A_{21}e_{22}}{d_1 s_1^2} = \frac{-[9.148 \times .135 + 664.903 \times .0175] \times 10^{-10}}{3.032 \times 10^{-3}} = -4.245 \times 10^{-7}$$

$$\Gamma_{22} = \frac{A_{12}e_{12} + A_{22}e_{22}}{d_1 s_2^2} = \frac{[976.753 \times .0175 - 31.272 \times .135] \times 10^{-10}}{4.455 \times 10^{-3}} = 2.889 \times 10^{-7}$$

$$L_{11} = \frac{B_{11}e_{11} + B_{21}e_{21}}{d_1 s_1^2} = \frac{-[25.157 \times 39.42 + 21.877 \times .135] \times 10^{-10}}{3.032 \times 10^{-3}} = -328.048 \times 10^{-7}$$

$$L_{12} = \frac{B_{12}e_{11} + B_{22}e_{21}}{d_1 s_2^2} = \frac{[36.956 \times 39.42 - 12.573 \times .135] \times 10^{-10}}{4.455 \times 10^{-3}} = 326.624 \times 10^{-7}$$

$$L_{21} = \frac{B_{11}e_{12} + B_{21}e_{22}}{d_1 s_1^2} = \frac{[+25.157 \times .135 + 21.877 \times .0175] \times 10^{-10}}{3.032 \times 10^{-3}} = 1.246 \times 10^{-7}$$

$$L_{22} = \frac{B_{12}e_{12} + B_{22}e_{22}}{d_1 s_2^2} = \frac{[-36.956 \times .135 + 12.573 \times .0175] \times 10^{-10}}{4.455 \times 10^{-3}} = -1.071 \times 10^{-7}$$

Also, let

$$F_{1_K}(t) = \begin{cases} 1 - \cos s_1 t & , \quad t < t_{0_K} \\ \qquad or \\ \cos s_1(t - t_{0_K}) - \cos s_1 t, & t \geq t_{0_K} \end{cases} \tag{83}$$

Where K takes the values 1, 2, 3,…, m for the m desired values of t_0

And, let:

$$G_{1_K}(t) = \begin{cases} s_1 t - \sin s_1 t & , \quad t < t_{0_K} \\ \qquad or \\ s_1 t \cos s_1(t - t_{0_K}) - \sin s_1 t - s_1(t - t_{0_K}) \cos s_1(t - t_{0_K}) + \sin s_1(t - t_{0_K}), & t \geq t_{0_K} \end{cases} \tag{84}$$

$$H_{1_K}(t) = \begin{cases} \dfrac{2\pi t_{0_K}}{\pi^2 - s_1^2 t_{0_K}^2}\sin s_1 t - \dfrac{2 s_1 t_{0_K}^2}{\pi^2 - s_1^2 t_{0_K}^2}\sin \dfrac{\pi t}{t_{0_K}} \quad , \quad t < t_{0_K} \\[4mm] \text{or} \\[2mm] \dfrac{2\pi t_{0_K}}{\pi^2 - s_1^2 t_{0_K}^2}\sin s_1 t - \dfrac{2\pi t_{0_K}}{\pi^2 - s_1^2 t_{0_K}^2}\Big[\cos s_1 t \sin s_1 t_{0_K} - \sin s_1 t \cos s_1 t_{0_K}\Big] \, t \ge t_{0_K} \end{cases} \tag{85}$$

{These functions have been obtained from Eqs. (64), (65), and the improved Eqs. (73$'$) and (73)"}

The response may be written with this notation as follows:

$$\begin{aligned}
\overset{\Diamond\,(1)}{Z_{M_K}}(t) &= \Gamma_{11}F_{1_K}(t) + \Gamma_{12}F_{1_{\frac{s2}{s1}K}}\left(\dfrac{s_2}{s_1}t\right) \\[2mm]
\overset{\Diamond\,(2)}{Z_{M_K}}(t) &= \Gamma_{21}F_{1_K}(t) + \Gamma_{22}F_{1_{\frac{s2}{s1}K}}\left(\dfrac{s_2}{s_1}t\right) \\[2mm]
\overset{\Diamond\,(1)}{\theta_{y_K}}(t) &= L_{11}F_{1_K}(t) + L_{12}F_{1_{\frac{s2}{s1}K}}\left(\dfrac{s_2}{s_1}t\right) \\[2mm]
\overset{\Diamond\,(2)}{\theta_{y_K}}(t) &= L_{21}F_{1_K}(t) + L_{22}F_{1_{\frac{s2}{s1}K}}\left(\dfrac{s_2}{s_1}t\right)
\end{aligned} \tag{86}$$

since $t_{0_K}^1 = \dfrac{s_2}{s_1}t_{0_K}$, must use (86), response for Type (a) function - square step

Referring to pages 32 through 34, the dynamic response factors may be written in terms of the maximax displacements obtained from Eqs. (86):

$$\overset{\Diamond}{\gamma}_{1_z} = \dfrac{\overset{\Diamond\,(1)}{Z_{Max}}}{Z_{M_{s(1)}}} = \dfrac{\overset{\Diamond\,(1)}{Z_{Max}}}{\alpha_{22}} = \dfrac{\overset{\Diamond\,(1)}{Z_{Max}}}{39.42}\times 10^6$$

$$\overset{\Diamond}{\gamma}_{2_z} = \dfrac{\overset{\Diamond\,(2)}{Z_{Max}}}{Z_{M_{s(2)}}} = \dfrac{\overset{\Diamond\,(2)}{Z_{Max}}}{\alpha_{24}} = \dfrac{\overset{\Diamond\,(2)}{Z_{Max}}}{.135}\times 10^6$$

(87) Dynamic response factors - Square step

$$\overset{\Diamond}{\gamma}_{1_\theta} = \dfrac{\overset{\Diamond\,(1)}{\theta_{Y Max}}}{\theta_{Y_{s(1)}}} = \dfrac{\overset{\Diamond\,(1)}{\theta_{Y Max}}}{\alpha_{42}} = \dfrac{\overset{\Diamond\,(1)}{\theta_{Y Max}}}{.135}\times 10^6$$

$$\overset{\Diamond}{\gamma}_{2_\theta} = \dfrac{\overset{\Diamond\,(2)}{\theta_{Y Max}}}{\theta_{Y_{s(2)}}} = \dfrac{\overset{\Diamond\,(2)}{\theta_{Y Max}}}{\alpha_{44}} = \dfrac{\overset{\Diamond\,(2)}{\theta_{Y Max}}}{.0175}\times 10^6$$

Also,

$$\overset{\vartriangle}{Z}{}_{M_K}^{(1)}(t) = \frac{\Gamma_{11}}{s_1 t_{0_K}} G_{1_K}(t) + \frac{\Gamma_{12}}{s_2 t_{0_K}} G_{1_{\frac{s_2}{s_1}K}}\left(\frac{s_2}{s_1}t\right)$$

$$\overset{\vartriangle}{Z}{}_{M_K}^{(2)}(t) = \frac{\Gamma_{21}}{s_1 t_{0_K}} G_{1_K}(t) + \frac{\Gamma_{22}}{s_2 t_{0_K}} G_{1_{\frac{s_2}{s_1}K}}\left(\frac{s_2}{s_1}t\right)$$

$$\overset{\vartriangle}{\theta}{}_{y_K}^{(1)}(t) = \frac{L_{11}}{s_1 t_{0_K}} G_{1_K}(t) + \frac{L_{12}}{s_2 t_{0_K}} G_{1_{\frac{s_2}{s_1}K}}\left(\frac{s_2}{s_1}t\right)$$

$$\overset{\vartriangle}{\theta}{}_{y_K}^{(2)}(t) = \frac{L_{21}}{s_1 t_{0_K}} G_{1_K}(t) + \frac{L_{22}}{s_2 t_{0_K}} G_{1_{\frac{s_2}{s_1}K}}\left(\frac{s_2}{s_1}t\right)$$

(88) Response for type (b) function – Ramp

The dynamic response factors given by Eq. (87) apply to the case also, where $\vartriangle$ replaces $\lozenge$.

And,

$$\overset{\frown}{Z}{}_{M_K}^{(1)}(t) = \frac{s_1}{2}\Gamma_{11} \cdot H_{1_K}(t) + \frac{s_2}{2}\Gamma_{12} \cdot H_{1_{\frac{s_2}{s_1}K}}\left(\frac{s_2}{s_1}t\right)$$

$$\overset{\frown}{Z}{}_{M_K}^{(2)}(t) = \frac{s_1}{2}\Gamma_{21} \cdot H_{1_K}(t) + \frac{s_2}{2}\Gamma_{22} \cdot H_{1_{\frac{s_2}{s_1}K}}\left(\frac{s_2}{s_1}t\right)$$

$$\overset{\frown}{\theta}{}_{y_K}^{(1)}(t) = \frac{s_1}{2}L_{11} \cdot H_{1_K}(t) + \frac{s_2}{2}L_{12} \cdot H_{1_{\frac{s_2}{s_1}K}}\left(\frac{s_2}{s_1}t\right)$$

$$\overset{\frown}{\theta}{}_{y_K}^{(2)}(t) = \frac{s_1}{2}L_{21} \cdot H_{1_K}(t) + \frac{s_2}{2}L_{22} \cdot H_{1_{\frac{s_2}{s_1}K}}\left(\frac{s_2}{s_1}t\right)$$

(89)

Response for type (c) function – Half sine

The dynamic response factors are again given by Eqs. (87), with $\frown$ replacing $\lozenge$.

Since the functions $F_{1_K}(t), G_{1_K}(t),$ and $H_{1_K}(t)$ are defined differently for $t < t_{0_K}$ and $t \geq t_{0_K}$, if we denote the function $F_{1_K}(t)$ valid for $t < t_{0_K}$ by $F_{1_K}(t)(t < t_{0_K})$, and the function $F_{1_K}(t)$ valid for $(t \geq t_{0_K})$ by $F_{1_K}(t)(t \geq t_{0_K})$, then the response defined by the first Eq. Of (86) can be written as:

$$\overset{\lozenge}{Z}{}_{M_K}^{(1)}(t < t_{0_K}) = \Gamma_{11}F_{1_K}(t < t_{0_K}) + \Gamma_{12}F_{1_{\frac{s_2}{s_1}K}}\left(\frac{s_2}{s_1}\left[t < t_{0_K}\right]\right)$$

$$\overset{\lozenge}{Z}{}_{M_K}^{(1)}(t \geq t_{0_K}) = \Gamma_{11}F_{1_K}(t \geq t_{0_K}) + \Gamma_{12}F_{1_{\frac{s_2}{s_1}K}}\left(\frac{s_2}{s_1}\left[t > t_{0_K}\right]\right)$$

(90)

The implication evident in these Eqs. is that the functions $F_{1_K}(t)(t < t_{0_K})$ and $F_{1_K}(t)(t \geq t_{0_K})$ must be

calculated out to arguments $F_{1_K}\left(\dfrac{S_2}{S_1}t_{0_K}\right)$ and $F_{1_K}\left(\dfrac{S_2}{S_1}t_{Max}\right)$, where t_{Max} is the maximum time for

which the response is desired. Of course, the same holds true for the functions G_{1_K} and H_{1_K}.

The response will be obtained for 15 values of t_0;

i.e., $\quad t_{0_K} = 0.1KT_1 = \dfrac{0.1K}{f_1} \quad ; K = 1,2,3,...,15$ $\hspace{3cm}$ (91)

Utilizing a digital computer, the results have been obtained and are presented in Appendix 4.

Dynamic response factors calculated from Eqs. (87) with the maximax values from the computer

results are given in Table VII.

2.9   Numerical Calculations – Y axis

Parameters:
$$\left.\begin{array}{l} \alpha_{11} = 38.57 \times 10^{-6} \\[2mm] \alpha_{15} = \dfrac{14.384}{216} \times 10^{-6} \\[2mm] \alpha_{55} = \dfrac{4.924}{108 \times 216} \times 10^{-6} \end{array}\right\} \quad \text{(Ref. Eq. (8) Appendix 2)}$$

$$\left.\begin{array}{l} I_{Z_M} = 532.8 \times 10^{3}, \quad M = 87.4 \\[2mm] I_{Z_R} = 10.0 \times 10^{6} \end{array}\right\} \quad \text{Ref. Page 41}$$

$$M = \frac{1}{4} M_{arm} + M_{Gon} + M_{Gim} + M_{Fork}$$
$$= 13.85 + 6.2 + 18.5 + 48.9$$
$$= 87.45 \quad lb \, sec^{2} \, in^{-1}$$

$$\left\{\begin{array}{l} a_{11} = \alpha_{11} M \\[2mm] = 33.57 \times 10^{-6} \times 87.45 \\[2mm] = 29.356965 \times 10^{-4} \\[4mm] \alpha_{12} = -\alpha_{15} I_{Z_M} \\[2mm] = -\dfrac{14.384}{216} \times 10^{-6} \times 52.38 \times 10^{4} \\[2mm] = -3.48812 \times 10^{-2} = -348.812 \times 10^{-4} \\[4mm] \left(\because I_{Z_M} = I_{Z_{Fork}} + I_{Z_{Gon}} + I_{Z_{Gim}} = 52.38 \times 10^{4} \right) \\[2mm] a_{13} = -600.00 \end{array}\right.$$

$$a_{21} = -\alpha_{15} M = \frac{-14.384}{216} \times 10^{-6} \times 87.45$$
$$= -5.823522 \times 10^{-6}$$
$$a_{22} = \alpha_{55} I_{Z_M} = \frac{4.924}{108 \times 216} \times 10^{-6} \times 52.34 \times 10^{4}$$
$$= \frac{4.924 \times 52.34 \times 10^{-2}}{2.3328 \times 10^{4}}$$
$$= 11.04776 \times 10^{-5}$$
$$a_{23} = -1$$

$$a_{31} = MR = 52.470 \times 10^3 \quad \longleftarrow$$

$$a_{32} = I_{Z_M} = 52.380 \times 10^4 \quad \longleftarrow$$

$$a_{33} = I_{Z_R} = I_{Z_{Tot}} - I_{Z_M} - MR^2$$

$$= 42.0 \times 10^6 - 52.38 \times 10^4 - (600)^2 \times 87.45$$

$$= [42.0000 - 0.5238 - 31.4820] \times 10^{+6}$$

$$= 9.9942 \times 10^{+6} \quad \longleftarrow$$

Characteristic determinant:

$$b_3 = a_{11}a_{22}a_{33} - a_{12}a_{21}a_{33}$$

$$= a_{33}(a_{11}a_{22} - a_{12}a_{21})$$

$$= 9.9942 \times 10^6 (29.3570 \times 11.0478 - 34.8812 \times 5.3235) \times 10^{-9}$$

$$= 9.9942 \times (324.529125 - 203.130668) \times 10^{-9+6}$$

$$= 1.2132 \quad \longleftarrow$$

$$b_2 = -(a_{11} + a_{22})a_{33} + a_{13}(a_{21}a_{32} + a_{22}a_{31}) + a_{23}(a_{12}a_{31} + a_{11}a_{32})$$

$$= -(293.5697 + 11.0477) \times 9.9942 \times 10^{6-5}$$

$$- (-5.8235 \times 5.2380 \times 10^{-6+5} + 11.0478 \times 5.247 \times 10^{-1}) \times 600$$

$$- (-3.4881 \times 5.2470 \times 10^{3-2} + 29.3578 \times 5.238 \times 10)$$

$$- (3.0444 + 0.1648 + 0.1355) \times 10^4$$

$$= -3.3447 \times 10^4 \quad \longleftarrow$$

$$b_1 = a_{33} - a_{13}a_{31} - a_{23}a_{32}$$

$$= 9.9942 \times 10^{+6} + 600 \times 52.47 \times 10^3 + 5.2380 \times 10^4$$

$$= [9.9942 + 31.4820 + 0.0524] \times 10^6$$

$$= 41.5286 \times 10^6$$

Conclusion:

$$1) \begin{bmatrix} M \\ I_{Z_M} \\ I_{Z_R} \end{bmatrix} = \begin{bmatrix} 87.45 \\ 52.38 \times 10^4 \\ 9.9942 \times 10^6 \end{bmatrix}$$

$$2) \begin{bmatrix} a_{11} & a_{12} & a_{13} \\ a_{21} & a_{22} & a_{23} \\ a_{31} & a_{32} & a_{33} \end{bmatrix} = \begin{bmatrix} 29.3570 \times 10^{-4} & -348.8120 \times 10^{-4} & -600.00 \\ -5.823 \times 10^{-6} & 110.4776 \times 10^{-6} & -1 \\ 52.470 \times 10^3 & 52.380 \times 10^4 & 9.9942 \times 10^6 \end{bmatrix}$$

$$3) \begin{bmatrix} b_1 \\ b_2 \\ b_3 \end{bmatrix} = \begin{bmatrix} 41.5286 \times 10^6 \\ -3.3447 \times 10^4 \\ 1.2132 \end{bmatrix}$$

$$4) \begin{bmatrix} s_1^2 \\ s_2^2 \end{bmatrix} = \begin{bmatrix} 1.3027 \\ 26.2663 \end{bmatrix} \times 10^3$$

$$A' = -\frac{b_2}{b_3} = \frac{3.3447}{1.2132} \times 10^4 = 2.7569 \times 10^4$$

$$B' = \frac{b_1}{b_3} = 34.2306 \times 10^6$$

$$\begin{bmatrix} A' \\ B' \end{bmatrix} = \begin{bmatrix} 2.7569 \times 10^4 \\ 34.2306 \times 10^6 \end{bmatrix}$$

$$\tau = -\frac{A'}{2} \pm \sqrt{\left(\frac{A'}{2}\right)^2 - B'}$$

$$= -1.3785 \times 10^4 \pm \sqrt{190.0262 - 34.2306} \times 10^3$$

$$= \left[-13.7845 \times 10^3 \pm \sqrt{155.7956} \times 10^3 \right]$$

$$= \left[-13.7845 \pm 12.4818 \right] \times 10^3$$

$$\begin{bmatrix} \tau_1 \\ \tau_2 \end{bmatrix} = - \begin{bmatrix} 1.3027 \\ 26.2663 \end{bmatrix} \times 10^3 \quad \longleftarrow$$

$$\begin{bmatrix} s_1^2 \\ s_2^2 \end{bmatrix} = + \begin{bmatrix} 1.3027 \\ 26.2663 \end{bmatrix} \times 10^3$$

$$s_2^2 - s_1^2 = 24.9636 \times 10^3$$

$$A_{12} = -\left[\frac{a_{22}a_{33}s_1^2 - a_{33} + a_{23}a_{32}}{s_2^2 - s_1^2} \right]$$

$$= -\left[\frac{a_{33}\left[a_{22}s_1^2 - 1\right] + a_{23}a_{32}}{24.9636 \times 10^3} \right]$$

$$= \frac{9.9942 \times 10^{+6} \left[+1.439 \times 10^{-1} - 1 \right] - 52.380 \times 10^4}{-24.9636 \times 10^3}$$

$$= \frac{8.560 \times 10^6 - 52.38 \times 10^4}{24.9636 \times 10^3}$$

$$= \frac{803.62 \times 10^1}{24.963} = 33.4842 \times 10$$

$$= 3.3484 \times 10^2$$

$$A_{13} = \frac{a_{33}(a_{22}s_2^2 - 1) + a_{23}a_{32}}{+24.9636 \times 10^3}$$

$$= \frac{9.9942 \times 10^6 [+2.9024 - 1] - 52.38 \times 10^4}{24.9636 \times 10^3}$$

$$= \frac{[+19.0129 - 0.5238] \times 10^6}{24.9636 \times 10^3}$$

$$= \frac{18.4891}{24.9636} \times 10^3$$

$$= 7.3956 \times 10^2$$

$$A_{21} = 0$$

$$A_{22} = \frac{a_{12}a_{33}s_1^2 - a_{13}a_{32}}{-24.9636 \times 10^3}$$

$$= \frac{-3.4861 \times 1.3027 \times 10^{3+5} + 3.1428 \times 10^8}{-24.9636 \times 10^3}$$

$$= \frac{-1.3985 \times 10^8}{-24.9636 \times 10^3}$$

$$= 5.5940 \times 10^3$$

$$A_{23} = \frac{a_{12}a_{33}s_2^2 - a_{13}a_{32}}{24.9936 \times 10^3}$$

$$= \frac{(-91.5669 + 3.1428) \times 10^5}{24.9935 \times 10^3}$$

$$= -3.5370 \times 10^5$$

$$A_{31} = \frac{-a_{13}}{s_1^2 s_2^2}$$

$$= \frac{-600}{1.3027 \times 26.2663 \times 10^6}$$

$$= \frac{-600}{34.2171 \times 10^{+6}}$$

$$= -1.7535 \times 10^{-5}$$

$$A_{32} = -A + \frac{B}{s_1^2}$$

$$= +40.4467 \times 10^{-8} + 1.8423 \times 10^{-5}$$

$$= -1.8029 \times 10^{-5}$$

$$A = \frac{\left[a_{12}a_{23} + a_{13}a_{22}\right]}{\eta}$$

$$= 40.4467 \times 10^{-8}$$

$$B = +\frac{a_{13}}{\eta}$$

$$= -2.4 \times 10^{-2}$$

$$A_{33} = A - \frac{B}{s_2^2}$$

$$= [-.40447 + 9.1372] \times 10^{-6}$$

$$= 8.7327 \times 10^{-6}$$

$$\begin{bmatrix} A_{11} & A_{12} & A_{13} \\ A_{21} & A_{22} & A_{23} \\ A_{31} & A_{32} & A_{33} \end{bmatrix} = \begin{bmatrix} 0 & 3.3484 \times 10^2 & 7.3956 \times 10^2 \\ 0 & 5.5940 \times 10^3 & -3.5370 \times 10^5 \\ 1.7535 \times 10^{-5} & -1.8029 \times 10^{-5} & 8.7327 \times 10^{-6} \end{bmatrix}$$

$$B_{11} = 0$$

$$B_{12} = \frac{-a_{21}a_{33}s_1^2 + a_{23}a_{31}}{s_2^2 - s_1^2}$$

$$= \frac{50.821 \times 1.3027 \times 10^3 - 5.247 \times 10^4}{2.5 \times 10^4} = \frac{(6.6205 - 5.2470) \times 10^4}{2.5 \times 10^4}$$

$$= 0.5494$$

$$B_{13} = \frac{a_{21}a_{33}s_2^2 - a_{23}a_{31}}{s_2^2 - s_1^2}$$

$$= \frac{-50.821 \times 26.2663 \times 10^3 + 5.247 \times 10^4}{2.5 \times 10^4} = \frac{-133.4880 + 5.2470}{2.5}$$

$$= -51.2964$$

$$B_{21} = 0$$

$$B_{22} = \frac{(a_{33} - a_{13}a_{31}) - a_{11}a_{33}s_1^2}{s_2^2 - s_1^2}$$

$$= \frac{[4.1476 - 2.9359 \times 1.3027] \times 10^7}{2.5 \times 10^4}$$

$$= 1.2920 \times 10^2$$

$$B_{23} = \frac{-(a_{33} - a_{13}a_{31}) + a_{11}a_{33}s_2^2}{s_2^2 - s_1^2}$$

$$= \frac{[-4.1476 + 2.9359 \times 26.2663] \times 10^3}{2.5}$$

$$= 2.9187 \times 10^4$$

$$B_{31} = -\frac{a_{23}}{s_1^2 s_2^2}$$

$$= \frac{-1}{3.4217 \times 10^7}$$

$$= -2.9225 \times 10^{-8}$$

$$B_{32} = \frac{-a_{11}a_{23} + a_{13}a_{21}}{2.5 \times 10^4} + \frac{a_{23}}{s_1^2(2.5 \times 10^4)}$$

$$= -2.2284 \times 10^{-9} - 3.0705 \times 10^{-8}$$

$$= -3.2933 \times 10^{-8}$$

$$B_{33} = 2.2284 \times 10^{-9} + \frac{a_{23}}{s_2^2(2.5 \times 10^4)}$$

$$= 2.2284 \times 10^{-9} - 1.5228 \times 10^{-9}$$

$$= 0.7056 \times 10^{-9}$$

$$\begin{bmatrix} B_{11} & B_{12} & B_{13} \\ B_{21} & B_{22} & B_{23} \\ B_{31} & B_{32} & B_{33} \end{bmatrix} = \begin{bmatrix} 0 & 0.5494 & -51.2964 \\ 0 & 1.2920 \times 10^2 & 2.9187 \times 10^4 \\ -2.9225 \times 10^{-8} & -3.2933 \times 10^{-8} & 7.056 \times 10^{-10} \end{bmatrix}$$

$$c_{11} = 0$$

$$c_{12} = \frac{s_1^2[a_{21}a_{32} + a_{22}a_{31}] - a_{31}}{s_2^2 - s_1^2} \qquad \left[\begin{array}{c} \alpha = a_{21}a_{32} + a_{22}a_{31} \\ \alpha = +2.4704 \end{array} \right]$$

$$= \frac{-2.4704 \times 1.3027 \times 10^3 - 52.4700 \times 10^{+3}}{2.5 \times 10^4}$$

$$= -2.2275$$

$$c_{13} = \frac{-s_2^2\alpha + a_{31}}{s_2^2 - s_1^2}$$

$$= \frac{[-2.4704 \times 26.2663 + 52.4700] \times 10^3}{2.5 \times 10^4}$$

$$= -0.4967$$

$$c_{21} = 0$$

$$c_{22} = \frac{s_1^2(a_{11}a_{32} + a_{12}a_{31}) - a_{32}}{\left(s_2^2 - s_1^2\right)} \qquad \left[\begin{array}{c} \beta = (a_{11}a_{32} + a_{12}aa_{32}) \\ = -25.7615 \end{array} \right]$$

$$= \frac{-26.7615 \times 1.3027 \times 10^3 - 5.238 \times 10^5}{2.5 \times 10^4} \qquad \left[s_2^2 - s_1^2 = S \right]$$

$$= -2.2346 \times 10$$

$$c_{23} = \frac{+a_{32} - \beta s_2^2}{S}$$

$$= \frac{+26.7615 \times 26.2663 \times 10^3 + 5.238 \times 10^5}{S}$$

$$= \frac{12.3743 \times 10^5}{2.5 \times 10^4}$$

$$= 49.4972$$

$$c_{31} = \frac{1}{s_1^2 s_2^2} = 2.9225 \times 10^{-8}$$

$$c_{32} = -s_1^2\gamma + \varepsilon + \frac{\sigma}{s_1^2} \qquad \left[\begin{array}{c} \alpha = \dfrac{a_{11}a_{22} - a_{12}a_{21}}{S} = 4.8492 \times 10^{-12} \\ \varepsilon = \dfrac{a_{11} + a_{22}}{S} = 1.2189 \times 10^{-7} \\ \sigma = \dfrac{1}{S} = 4 \times 10^{-5} \end{array} \right]$$

$$= -6.3170 \times 10^{-9} + 1.2189 \times 10^{-7} + 3.0705 \times 10^{-8}$$

$$= 1.4628 \times 10^{-7}$$

$$c_{33} = s_2^2 \gamma - \varepsilon + \frac{\sigma}{s_2^2}$$
$$= 1.2737 \times 10^{-7} - 1.2189 \times 10^{-7} + 1.0238 \times 10^{-9}$$
$$= 6.5038 \times 10^{-9}$$

$$\begin{bmatrix} c_{11} & c_{12} & c_{13} \\ c_{21} & c_{22} & c_{23} \\ c_{31} & c_{32} & c_{33} \end{bmatrix} = \begin{bmatrix} 0 & -2.2275 & -0.4967 \\ 0 & -22.3460 & 49.4972 \\ 2.9225 \times 10^{-8} & 1.4628 \times 10^{-7} & 6.5038 \times 10^{-7} \end{bmatrix}$$

Conclusions:

$$\begin{bmatrix} A_{11} & A_{12} & A_{13} \\ A_{21} & A_{22} & A_{23} \\ A_{31} & A_{32} & A_{33} \end{bmatrix} = \begin{bmatrix} 0 & 3.3484 \times 10^2 & 7.3956 \times 10^2 \\ 0 & 5.5940 \times 10^3 & -3.5370 \times 10^5 \\ 1.7535 \times 10^{-5} & -1.8029 \times 10^{-5} & 8.7327 \times 10^{-6} \end{bmatrix}$$

$$\begin{bmatrix} B_{11} & B_{12} & B_{13} \\ B_{21} & B_{22} & B_{23} \\ B_{31} & B_{32} & B_{33} \end{bmatrix} = \begin{bmatrix} 0 & 0.5494 & -51.2964 \\ 0 & 1.2920 \times 10^2 & 2.9187 \times 10^4 \\ -2.9225 \times 10^{-8} & -3.2933 \times 10^{-8} & 7.056 \times 10^{-10} \end{bmatrix}$$

$$\begin{bmatrix} c_{11} & c_{12} & c_{13} \\ c_{21} & c_{22} & c_{23} \\ c_{31} & c_{32} & c_{33} \end{bmatrix} = \begin{bmatrix} 0 & -2.2275 & -0.4967 \\ 0 & -22.3460 & 49.4972 \\ 2.9225 \times 10^{-8} & 1.4628 \times 10^{-7} & 6.5038 \times 10^{-7} \end{bmatrix}$$

$$[D_{i_j}] = \begin{bmatrix} 0 & 16.7134 \times 10^2 & 10.3758 \times 10^2 \\ 0 & 1.9002 \times 10^2 & -3.8340 \times 10^5 \\ 0 & -1.0576 \times 10^{-4} & 3.8149 \times 10^{-4} \end{bmatrix}$$

$$[E_{i_j}] = \begin{bmatrix} 0 & 2.7769 & -50.7997 \\ 0 & 1.5155 \times 10^2 & 291.3800 \times 10^2 \\ 0 & -1.7921 \times 10^{-7} & -6.5109 \times 10^{-7} \end{bmatrix}$$

$$K_1 = MR^2 + I_{Z_M} + I_{Z_R} = I_{Z_{Tot}}$$
$$= 42.0000 \times 10^6$$

$$K_2 = \alpha_{11} MR + \alpha_{15} I_{Z_M}$$
$$= 17.6114 + 0.3488$$
$$= 17.6498$$

$$K_3 = \alpha_{15} MR + \alpha_{55} I_{Z_M}$$
$$= [3.4941 + 11.0467] \times 10^{-5} = 14.5408 \times 10^{-5}$$

$$\begin{bmatrix} K_1 \\ K_2 \\ K_3 \end{bmatrix} = \begin{bmatrix} 42.0000 \times 10^6 \\ 17.6498 \\ 14.5408 \times 10^{-5} \end{bmatrix}$$

$$[e_{i_j}] = \begin{bmatrix} 1 & R & 1 \\ 0 & \alpha_{11} & \alpha_{15} \\ 0 & \alpha_{15} & \alpha_{55} \end{bmatrix} = \begin{bmatrix} 1 & 600 & 1 \\ 0 & 33.57 \times 10^{-6} & 6.6593 \times 10^{-7} \\ 0 & 6.6593 \times 10^{-7} & 2.1108 \times 10^{-5} \end{bmatrix}$$

From Page 22;

$$\underline{D_{i_j} = [A_{ij}] - R[C_{ij}]}$$

Solution

$$R[C_{ij}] = \begin{bmatrix} 0 & -13.3650 \times 10^2 & -2.9802 \times 10^2 \\ 0 & -134.0760 \times 10^2 & 296.9832 \times 10^2 \\ 1.7535 \times 10^{-5} & 8.7768 \times 10^{-5} & 39.0228 \times 10^{-5} \end{bmatrix}$$

$$[D_{ij}] = \begin{bmatrix} 0 & (3.3484 + 13.3650) \times 10^2 & (7.3956 + 2.9802) \times 10^2 \\ 0 & (5.5840 + 13.4076) \times 10^3 & (-3.5370 - 0.2970) \times 10^5 \\ 0 & (-1.8029 - 8.7768) \times 10^{-5} & (8.7327 - 390.2280) \times 10^{-6} \end{bmatrix}$$

$$= \begin{bmatrix} 0 & 16.7134 \times 10^2 & 10.3758 \times 10^2 \\ 0 & 19.0016 \times 10^3 & -3.8340 \times 10^5 \\ 0 & -10.5759 \times 10^{-5} & 381.4953 \times 10^{-6} \end{bmatrix}$$

$$[E_{i_j}] = [B_{ij}] - [C_{ij}]$$

$$= \begin{bmatrix} 0 & (0.5494 + 2.2275) & (-51.2964 + 0.4967) \\ 0 & (1.2920 + .2235)10^2 & (2.9187 - 0.0049)10^4 \\ 0 & (-3.2933 - 14.6280)10^{-8} & (-6.5038 + 0.0071)10^{-7} \end{bmatrix}$$

$$= \begin{bmatrix} 0 & 2.7769 & -50.7997 \\ 0 & 1.5155 \times 10^2 & 2.9138 \times 10^4 \\ 0 & -17.9213 \times 10^{-8} & -6.5109 \times 10^{-7} \end{bmatrix} \Leftarrow$$

CHAPTER 3 Gimbal Response

3.1 Overview

A model of the centrifuge system will be studied in the vertical plane in order to determine dynamic load factors applicable to the gimbal ring. The results of the analysis will be employed in appendix 6 in order to justify the uncoupling of the system tacitly assumed possible in Chapter 1. The dynamic response of the system in the vertical plane with the gimbal ring in $\theta = 0°$ position will be investigated; it is felt that the gimbal ring is considerably stiffer in the other planes and that, as a result, the current study will lead to results which may be applied to the other planes.

In this analysis the gimbal ring will be idealized as a simple beam with the gondola mass at its center. Using Rayleigh's method, the beam will be replaced by a simple spring with one half of the ring mass and the gondola mass acting at the end.

The arm will be idealized as a simple cantilever beam, carrying one quarter of its mass at the tip.

3.2 Analysis

3.2.1 Model of System in the Vertical plane

$$m_1 = m_{Forks} + 0.25m_{Arm} + 0.50m_{Gimbal}$$
$$m_2 = m_{Gondola} + 0.50m_{Gimbal}$$

IDEALIZED SYSTEM

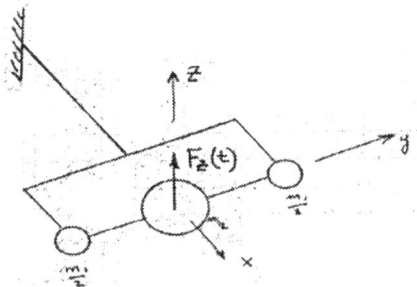

FIGURE 3.2.1

MATHEMATICAL MODEL

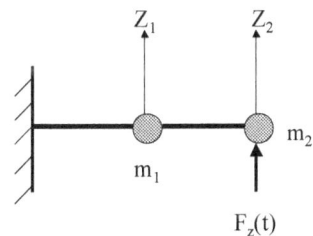

FIGURE 3.2.2

3.2.2 Analysis in the Vertical Plane

Initial Conditions:

$$Z_{1_0} = \dot{Z}_{1_0} = Z_{Z_0} = \dot{Z}_{Z_0} = 0$$

Equations of Motion:

$$\begin{cases} Z_1 = \alpha_{11}(-m_1\ddot{Z}_1) + \alpha_{12}(-m_2\ddot{Z}_2) + \alpha_{12}F_Z(t) \\ Z_2 = \alpha_{21}(-m_1\ddot{Z}_1) + \alpha_{22}(-m_2\ddot{Z}_2) + \alpha_{22}F_Z(t) \end{cases} \tag{1}$$

Solution for Transient Response

$$\begin{cases} u_1 = -\alpha_{11}m_1 s^2 u_1 - \alpha_{12}m_2 s^2 u_2 + \alpha_{12}f(s) \\ u_2 = -\alpha_{21}m_1 s^2 u_1 - \alpha_{22}m_2 s^2 u_2 + \alpha_{22}f(s) \end{cases} \qquad (2)$$

Where s = Laplacian operator = d/dt

$$u_1(t) = L\{Z_1(t)\}$$
$$u_2(t) = L\{Z_2(t)\}$$
$$f(s) = L\{F_Z(t)\}$$

Hence:

$$[s^2 A_0 + A_1][u] = [A_2]f(s) \qquad (3)$$

Where

$$[A_0] = \begin{bmatrix} \alpha_{11}m_1 & \alpha_{12}m_2 \\ \alpha_{21}m_1 & \alpha_{22}m_2 \end{bmatrix}; \qquad [u] = \begin{bmatrix} u_1 \\ u_2 \end{bmatrix}$$

$$[A_1] = \begin{bmatrix} 1 & 0 \\ 0 & 1 \end{bmatrix}; \qquad [A_2] = \begin{bmatrix} \alpha_{12} \\ \alpha_{22} \end{bmatrix}$$

$$[A_3] = [s^2 A_0 + A_1] = \begin{bmatrix} s^2\alpha_{11}m_1 + 1 & s^2\alpha_{12}m_2 \\ s^2\alpha_{21}m_1 & s^2\alpha_{22}m_2 + 1 \end{bmatrix}$$

The characteristic equation is:

$$|A_3| = (s^2\alpha_{11}m_1 + 1)(s^2\alpha_{22}m_2 + 1) - s^4\alpha_{12}\alpha_{21}m_1 m_2 = p(s)$$
$$\therefore \quad p(s) = s^4(\alpha_{11}\alpha_{22} - \alpha_{12}\alpha_{21})m_1 m_2 + s^2(\alpha_{11}m_1 + \alpha_{22}m_2) + 1$$

Setting $\tau = s^2$, the roots of the characteristic equation are given from $|A_3| = 0$:

$$\{\tau^2 + b_1\tau + b_2 = 0\} \qquad (4)$$

Where:

$$b_1 = \frac{\alpha_{11}m_1 + \alpha_{22}m_2}{\alpha_{11}\alpha_{22} - \alpha_{12}\alpha_{21}} - \frac{1}{m_1 m_2}$$

$$b_2 = \frac{1}{\alpha_{11}\alpha_{22} - \alpha_{12}\alpha_{21}} - \frac{1}{m_1 m_2}$$

From Eq. (4):

$$\tau = -\frac{b_1}{2} \pm \sqrt{\left(\frac{b_1}{2}\right)^2 - b_2} \qquad (5)$$

Let $\tau_1 < \tau_2$ and let $s_1^2 = -\tau_1$; $s_2^2 = -\tau_2$,

Then, the characteristic Eq. can be written in terms of roots as:

$$p(s) = \frac{1}{b_2}\left(s^2 + s_1^2\right)\left(s^2 + s_2^2\right)$$

Now, solving Eq. (3) using Cramer's rule, we obtain:

$$u_1 = \frac{\begin{vmatrix} \alpha_{12}f(s) & s^2\alpha_{12}m_2 \\ \alpha_{22}f(s) & s^2\alpha_{22}m_2+1 \end{vmatrix}}{p(s)} = b_2 \frac{\alpha_{12}\left(s^2\alpha_{22}m_2+1\right)-\alpha_{22}\left(s^2\alpha_{12}m_2\right)}{\left(s^2+s_1^2\right)\left(s^2+s_2^2\right)}f(s) \tag{6}$$

$$u_2 = \frac{\begin{vmatrix} s^2\alpha_{11}m_1+1 & \alpha_{12}f(s) \\ s^2\alpha_{21}m_1 & \alpha_{22}f(s) \end{vmatrix}}{p(s)} = b_2 \frac{\alpha_{12}\left(s^2\alpha_{11}m_1+1\right)-\alpha_{12}\left(s^2\alpha_{21}m_1\right)}{\left(s^2+s_1^2\right)\left(s^2+s_2^2\right)}f(s) \tag{7}$$

Expanding Eqs. (6) and (7) by partial fractions, we have:

$$u_1 = b_2 f(s)\left\{\frac{A_{11}}{s^2+s_1^2} + \frac{A_{12}}{s^2+s_2^2}\right\} \tag{8}$$

$$u_2 = b_2 f(s)\left\{\frac{B_{11}}{s^2+s_1^2} + \frac{B_{12}}{s^2+s_2^2}\right\} \tag{9}$$

Where:

$$\begin{aligned} A_{11}(s^2+s_2^2) + A_{12}(s^2+s_1^2) &= \alpha_{12}(s^2\alpha_{22}m_2+1) - \alpha_{22}(s^2\alpha_{12}m_2) \\ B_{11}(s^2+s_2^2) + B_{12}(s^2+s_1^2) &= \alpha_{22}(s^2\alpha_{11}m_1+1) - \alpha_{12}(s^2\alpha_{21}m_1) \end{aligned} \tag{10}$$

Substituting $s^2 = -s_1^2$ and $-s_2^2$ successively into Eqs. (10):

$$A_{11}(s_2^2-s_1^2) = \alpha_{12}(-s_1^2\alpha_{22}m_2+1) + \alpha_{22}s_1^2\alpha_{12}m_2$$

$$A_{11} = \frac{+\alpha_{12}\alpha_{22}s_1^2m_2 - \alpha_{12}\alpha_{22}s_1^2m_2 + \alpha_{12}}{s_2^2-s_1^2} = \frac{\alpha_{12}}{s_2^2-s_1^2}$$

$$A_{12}(s_1^2-s_2^2) = \alpha_{12}$$

$$A_{12} = -\frac{\alpha_{12}}{s_2^2-s_1^2}$$

$$B_{11}(s_2^2 - s_1^2) = \alpha_{22}(-s_1^2\alpha_{11}m_1 + 1) + \alpha_{12}\alpha_{21}s_1^2 m_1$$

$$B_{11} = \frac{s_1^2 m_1(\alpha_{12}\alpha_{21} - \alpha_{11}\alpha_{22}) + \alpha_{22}}{s_2^2 - s_1^2}$$

$$B_{12}(s_2^2 - s_1^2) = \alpha_{22}(-s_2^2\alpha_{11}m_1 + 1) + \alpha_{12}\alpha_{21}s_2^2 m_1$$

$$B_{12} = \frac{s_2^2 m_1(\alpha_{11}\alpha_{22} - \alpha_{12}\alpha_{21}) - \alpha_{22}}{s_2^2 - s_1^2}$$

The inverse transformation of Eqs. (8) and (9) may now be taken using the convolution theorem, since

$$L^{-1}\{f(s) \cdot g(s)\} = \int_0^t F(\beta)G(t - \beta)\, d\beta$$

where $\qquad L^{-1}\{f(s)\} = F_Z(t), \qquad L^{-1}\{g(s)\} = G(t)$

$$L^{-1}\left\{\frac{A_{ij} \cdot f(s)}{s^2 + s_j^2}\right\} = \frac{A_{ij}}{s_j}\int_0^t F_Z(\beta)\sin s_j(t - \beta)\, d\beta \tag{11}$$

Also, $L^{-1}\{u_1(s)\} = Z_1(t), \qquad L^{-1}\{u_2(s)\} = Z_2(t)$

The solution for the transient response is given as follows:

$$\begin{Bmatrix} Z_1(t) \\ Z_2(t) \end{Bmatrix} = b_2 \sum_{j=1}^{2} \begin{Bmatrix} A_{ij} \\ B_{ij} \end{Bmatrix} L^{-1}\left[\frac{f(s)}{s^2 + s_j^2}\right] \tag{12}$$

Restricting our investigations to the function already used in the analysis for the arm response (i.e., square step, ramp, and half sine loading types) we have from Eqs. (39), (40), (73)', (73)"; and Eq. (12):

Type (a) (square step):

$$\begin{Bmatrix} Z_1(t) \\ Z_2(t) \end{Bmatrix} = b_2 F_{Z_{Max}} \sum_{j=1}^{2} \begin{Bmatrix} A_{ij} \\ B_{ij} \end{Bmatrix} \begin{bmatrix} \begin{cases} \dfrac{1}{s_j^2}[1 - \cos s_j t] \quad, \quad t < t_0 \\ \qquad\qquad or \\ \dfrac{1}{s_j^2}[\cos s_j(t - t_0) - \cos s_j t] \quad, \quad t \geq t_0 \end{cases} \end{bmatrix} \tag{13}$$

Type (b) (ramp):

$$
\begin{Bmatrix} Z_1(t) \\ Z_2(t) \end{Bmatrix} = b_2 F_{Z_{Max}} \left[\sum_{j=1}^{2} \begin{Bmatrix} A_{i_j} \\ B_{i_j} \end{Bmatrix} \begin{Bmatrix} s_j t - \sin s_j t \quad , \quad t < t_0 \\ or \\ s_j t \cos s_j (t - t_0) - \sin s_j t - s_j (t - t_0) \cos s_j (t - t_0) \\ + \sin s_j (t - t_0), \quad t \geq t_0 \end{Bmatrix} \cdot \frac{1}{s_j^3 t_0} \right] \quad (14)
$$

Type (c) (half sine):

$$
\begin{Bmatrix} Z_1(t) \\ Z_2(t) \end{Bmatrix} = b_2 F_{Z_{Max}} \left[\sum_{j=1}^{2} \begin{Bmatrix} A_{i_j} \\ B_{i_j} \end{Bmatrix} \begin{Bmatrix} \dfrac{2\pi t_0}{\pi^2 - s_j^2 t_0^2} \sin s_j t - \dfrac{2 s_j t_0^2}{\pi^2 - s_j^2 t_0^2} \sin \dfrac{\pi t}{t_0} \quad , \quad t < t_0 \\ or \\ \dfrac{2\pi t_0}{\pi^2 - s_j^2 t_0^2} \sin s_j t \\ - \dfrac{2\pi t_0}{\pi^2 - s_j^2 t_0^2} [\sin s_j t_0 \cos s_j t - \cos s_j t_0 \sin s_j t_0], \quad t \geq t_0 \end{Bmatrix} \cdot \frac{1}{2 s_j} \right]
$$

$$(15)$$

3.2.3 Dynamic Response Factors

The static deflections under the peak transient loading $F_{Z_{Max}}$ are, from Eqs. (1):

$$
\begin{cases} Z_{1_S} = \alpha_{12} F_{Z_{Max}} \\ Z_{2_S} = \alpha_{22} F_{Z_{Max}} \end{cases}
$$

Thus, for unit values of the peak loading, we have:

$$
\begin{matrix} F_{Z_{Max}} = 1 \\ j = 1 \end{matrix} \quad \begin{cases} Z_{1_S} = \alpha_{12} \\ Z_{2_S} = \alpha_{22} \end{cases}
$$

From there static deflections and the transient displacements defined by Eqs. (13), (14), & (15), dynamic response factors in the form.

$$
\begin{cases} \gamma_{j_{Z_1}} = Z_{Max_1}^{j} / Z_{1_S} \\ \gamma_{j_{Z_2}} = Z_{Max_1}^{j} / Z_{2_S} \end{cases} \quad j = 1
$$

may be calculated

However, of particular interest is the force existing between masses 1 and 2; i.e., the dynamic spring force.

$$F_{spring} = k_2(Z_2 - Z_1) \qquad \text{where} \qquad k_2 = \frac{1}{\Delta}\{ \text{ from appendix 5}$$

Thus, from Eqs. (12) we obtain

$$F_{spring} = k_2 b_2 \sum_{j=1}^{2} (B_{1_j} - A_{1_j}) L^{-1}\left[\frac{f(s)}{s^2 + s_j^2}\right]$$

$$= k_2 b_2 F_{Z_{Max}} \sum_{j=1}^{2} (B_{1_j} - A_{1_j})\left\{ \quad \right\} \times (\text{constant})_j \qquad (16)$$

If we define the dynamic load factor as the ratio of the peak spring force to the peak loading, we have

$$\gamma_{dlf} = \frac{F_{spring}^{Max}}{F_{Z_{Max}}} = k_2 b_2 \sum_{j=1}^{2} (B_{1_j} - A_{1_j})\left\{\left\{ \quad \right\}\right\} \times (\text{constant})_j \qquad (17)$$

3.3 Numerical Calculations

$$\left.\begin{array}{l} \alpha_{11} = 39.42 \times 10^{-6} \\ \alpha_{12} = \alpha_{21} = \alpha_{11} \\ \alpha_{22} = 41.54 \times 10^{-6} \end{array}\right\} \qquad \text{(Ref. Appendix 5)}$$

$$\left.\begin{array}{l} m_1 = m_{Forks} + 0.25 m_{Arm} + 0.50 m_{Gimbal} = 44.50 \\ m_2 = m_{Gondola} + 0.50 m_{Gimbal} = 42.95 \end{array}\right\} \qquad \text{(Ref. pg. 41 Chapter 1)}$$

The coefficients of pg. 4 are:

$$b_1 = \frac{\alpha_{11} m_1 + \alpha_{22} m_2}{\alpha_{11}\alpha_{22} - \alpha_{12}\alpha_{21}} \cdot \frac{1}{m_1 m_2} = 22.153 \times 10^3$$

$$b_2 = \frac{1}{\alpha_{11}\alpha_{22} - \alpha_{12}\alpha_{21}} \cdot \frac{1}{m_1 m_2} = 6.2608 \times 10^6$$

Hence, from Eq. (5):

$$s_1^2 = \frac{b_1}{2} - \sqrt{\left(\frac{b_1}{2}\right)^2 - b_2} = 1.1076 \times 10^4 - 1.0790 \times 10^4 = 286.0$$

$$s_2^2 = \frac{b_2}{2} + \sqrt{\left(\frac{b_1}{2}\right)^2 - b_2} = 1.1076 \times 10^4 + 1.0790 \times 10^4 = 21,866.$$

$$A_{11} = -A_{12} = \frac{\alpha_{12}}{s_2^2 - s_1^2} = 1.8267 \times 10^{-9}$$

$$B_{11} = \frac{-s_1^2 m_1 (\alpha_{11}\alpha_{22} - \alpha_{12}\alpha_{21}) + \alpha_{22}}{s_2^2 - s_1^2} = 1.8756 \times 10^{-9}$$

$$B_{12} = \frac{s_2^2 m_1 (\alpha_{11}\alpha_{22} - \alpha_{12}\alpha_{21}) - \alpha_{22}}{s_2^2 - s_1^2} = 1.8432 \times 10^{-9}$$

$$s_1 = 16.91 \qquad \text{Rad/sec.}$$

$$s_2 = 147.87 \qquad \text{Rad/sec.}$$

The procedure followed in Chapter 1 for the transient response of the arm in the Z plane will be used here also. The digital computer program described in Appendix 4 will be used to obtain the response to the three types of forcing functions given in Eqs. (13), (14), & (15). Using the notation and method of Pg. 55, Chapter 1:

$$\Gamma_{11} = b_2 \frac{A_{11}}{s_1^2} = \frac{6.2608 \times 10^6 \times 1.8267 \times 10^{-9}}{2.86 \times 10^2} = 3.99881 \times 10^{-5}$$

$$\Gamma_{12} = b_2 \frac{A_{12}}{s_2^2} = \frac{6.2608 \times 10^6 \times (-1.8267 \times 10^{-9})}{21.866 \times 10^3} = -0.52303 \times 10^{-6}$$

$$L_{11} = b_2 \frac{B_{11}}{s_1^2} = \frac{6.2608 \times 10^6 \times 1.8756 \times 10^{-9}}{2.86 \times 10^2} = 4.10585 \times 10^{-5}$$

$$L_{12} = b_2 \frac{B_{12}}{s_2^2} = \frac{6.2608 \times 10^6 \times 1.8432 \times 10^{-9}}{21.866 \times 10^3} = 0.52775 \times 10^{-6}$$

The deformation of the gimbal spring can be obtained as follows:

From Eqs. (13):

$$\left\{ \begin{array}{l} \overset{\lozenge\ (1)}{Z_{1K}}(t) = \Gamma_{11} F_{1_K}(t) + \Gamma_{12} F_{1_{\frac{s_2}{s_1}K}}\left(\frac{s_2}{s_1}t\right) \\[2ex] \overset{\lozenge\ (1)}{Z_{2K}}(t) = L_{11} F_{1_K}(t) + L_{12} F_{1_{\frac{s_2}{s_1}K}}\left(\frac{s_2}{s_1}t\right) \end{array} \right\} \tag{18}$$

Hence, $\overset{\lozenge\ (1)}{Z_{2K}}(t) - \overset{\lozenge\ (1)}{Z_{1K}}(t) = (L_{11} - \Gamma_{11}) F_{1_K}(t) + (L_{12} - \Gamma_{12}) F_{1_{\frac{s_2}{s_1}K}}\left(\frac{s_2}{s_1}t\right)$

Let Y(t)= $Z_{2K}(t) - Z_{1K}(t)$ = spring deflection

And

$$\Gamma_{11}' = L_{11} - \Gamma_{11} = 1.07040 \times 10^{-6}$$

$$\Gamma_{12}' = L_{12} - \Gamma_{12} = 1.05078 \times 10^{-6}$$

Similarly, $$\overset{\lozenge}{Y}(t) = \Gamma_{11}' F_{1_K}(t) + \Gamma_{12}' F_{1_{\frac{s_2}{s_1}K}}\left(\frac{s_2}{s_1}t\right) \tag{19}$$

$$\overset{\lozenge}{Y}(t) = \frac{1}{s_1 t_{0_K}} \Gamma_{11}' G_{1_K}(t) + \frac{1}{s_2 t_{0_K}} \Gamma_{12}' G_{1_{\frac{s_2}{s_1}K}}\left(\frac{s_2}{s_1}t\right) \tag{20}$$

And $$\overset{\lozenge}{Y}(t) = \frac{s_1}{2} \Gamma_{11}' H_{1_K}(t) + \frac{s_2}{2} \Gamma_{12}' H_{1_{\frac{s_2}{s_1}K}}\left(\frac{s_2}{s_1}t\right) \tag{21}$$

Now, From Eq. (17) we have

$$\gamma_{dlf} = \frac{F_{Spring}^{Max}}{F_{Z_{Max}}} = \frac{k_2(Z_2 - Z_1)_{Max}}{F_{Z_{Max}}} \qquad\qquad \left\{ k_2 = \frac{10^6}{2.116} = \frac{1}{\Delta} \right.$$

$$\therefore \gamma_{dlf} = \frac{10^6}{2.116} \cdot Y_{Max} = \frac{Y'_{Max}}{2.116} \qquad\qquad\qquad \text{Ref. Appendix 5}$$

Where Y'_{Max} can be obtained from the computer results for each of the loading types used.[4]

Table IX, page 12, tabulates the values of γ_{dlf} for each of the loading cases and for each of the values of t_0.

---

[4] See Appendix 4

CHAPTER 4   Forcing Functions Arising From Rigid Body Kinematics

4.1 Overview

From the kinematic analysis of rigid body loads acting on the gimbal and the main arm, transient loads acting on the system may be obtained. Also, harmonic forcing functions resulting from steady application of angular velocity about the 1st and 2nd axes may be obtained.

This Chapter of the structural dynamics analysis derives the transient and steady state forcing functions acting on the system from the kinematic equations. The restraints on the physical system, expressed in terms of maximum rotational speeds and accelerations, are used in conjunction with dynamic response factors for the simple oscillator in order to reduce complexity of the loading equations wherever possible. It is shown that terms involving velocities are not required in the transient load equations, and that the response to transient forces which include such terms in static.

It is also shown that the harmonic forcing functions are also non-critical from a structural dynamic standpoint; i.e., static response occurs to the low frequency harmonic forces.

4.2 Rigid Body Kinematic Loads Acting on The Main Arm

From the rigid body kinematic analysis, forces and moments acting at the end of the arm may be obtained. The Eqs., simplified to let the principal axes of the gondola coincide with the axes of rotation, are as follows:

$$\begin{cases} r \Rightarrow gimbal \quad ring \\ g \Rightarrow gondola \end{cases}$$

$$B_{U_0} = -m_g \ddot{\psi} R$$
$$= -m_r \{ \ddot{\psi} R + \dot{\psi}^2 \bar{y}_r + 2\dot{\theta}\dot{\psi}(\bar{x}_r \sin\theta - \bar{z}_r \cos\theta) \\ - \ddot{\psi}(\bar{x}_r \cos\theta + \bar{z}_r \sin\theta) \}$$

$$A_{U_0} = -m_g G - m_r \{ G + \dot{\theta}^2 (\bar{x}_r \sin\theta - \bar{z}_r \cos\theta) \\ - \ddot{\theta}(\bar{x}_r \cos\theta + \bar{z}_r \sin\theta) \}$$

$$M_{A_U} = I_{x_g} \left[\dot{\theta}\dot{\phi}\cos\theta - (2\dot{\theta}\dot{\psi}\cos\theta + \ddot{\psi}\sin\theta - \ddot{\phi})\sin\theta \right] - \left[(\ddot{\psi}\cos\theta - 2\dot{\theta}\dot{\psi}\sin\theta)\cos\theta \right] \left(\frac{I_{y_g} + I_{z_g}}{2} \right)$$

$$+ \left[\ddot{\psi}\cos^2\theta - \dot{\theta}(\dot{\psi}\sin 2\theta + 2\dot{\phi}\cos\theta) \right] \left(\frac{I_{y_g} - I_{z_g}}{2} \right) \cos 2\phi$$

$$+ \left[\dot{\theta}^2 \sin\theta - (\ddot{\theta} + 2\dot{\phi}\dot{\psi}\cos\theta)\cos\theta \right] \left(\frac{I_{y_g} - I_{z_g}}{2} \right) \sin 2\phi$$

$$+ m_r \left[\ddot{\psi}R(\bar{x}_r \cos\theta + \bar{z}_r \sin\theta) + \dot{\psi}^2 R\bar{y}_r \right] - \ddot{\psi} \left[\frac{I_{x_r} + I_{z_r}}{2} - \frac{I_{x_r} - I_{z_r}}{2}\cos 2\theta + I_{xz_r}\sin\theta \right]$$

$$- \dot{\theta}\dot{\psi} \left[(I_{x_r} - I_{z_r})\sin 2\theta + 2I_{xz_r}\cos 2\theta \right] - \dot{\theta}^2 \left[I_{xy_r}\cos\theta + I_{yz_r}\sin\theta \right]$$

$$- \ddot{\theta} \left[I_{xy_r}\sin\theta - I_{yz_r}\cos\theta \right]$$

$$M_{B_U} = I_{x_g}\left[\left(\ddot{\phi} - \dot{\psi}\sin\theta\right)\dot{\psi}\cos\theta\right] + \left[\ddot{\theta} + \dot{\psi}^2\frac{\sin 2\theta}{2}\right]\frac{I_{y_g} + I_{z_g}}{2}$$

$$+ \left[\ddot{\theta} + \left(2\dot{\phi} - \dot{\psi}\sin\theta\right)\dot{\psi}\cos\theta\right]\left(\frac{I_{y_g} - I_{z_g}}{2}\cos 2\phi\right) + \left[\ddot{\psi}\cos\theta - 2\dot{\theta}\dot{\phi}\right]\left(\frac{I_{y_g} - I_{z_g}}{2}\sin 2\phi\right)$$

$$- m_r\left[\dot{\psi}^2 R(\bar{x}_r\sin\theta - \bar{z}_r\cos\theta) + G(\bar{x}_r\cos\theta + \bar{z}_r\sin\theta)\right] + \ddot{\theta}I_{y_r}$$

$$- \dot{\psi}^2\left[\frac{I_{x_r} - I_{z_r}}{2}\sin 2\theta + I_{xz_r}\cos 2\theta\right] + \ddot{\psi}\left[I_{xy_r}\sin\theta - I_{yz_r}\cos\theta\right]$$

$$M_{R_U} = I_{x_g}\left[\dot{\theta}\left(\dot{\psi}\cos 2\theta + \dot{\phi}\sin\theta\right) + \left(\ddot{\psi}\sin\theta - \ddot{\phi}\right)\cos\theta\right] - \left[\left(\ddot{\psi}\cos\theta - 2\dot{\theta}\sin\theta\right)\sin\theta\right]\frac{I_{y_g} + I_{z_g}}{2}$$

$$+ \left[\left(\ddot{\psi}\cos\theta - 2\dot{\theta}\dot{\phi}\right)\sin\theta + 2\dot{\theta}\dot{\psi}\cos^2\theta\right]\left[\frac{I_{y_g} - I_{z_g}}{2}\cos 2\phi\right]$$

$$+ \left[\left(\dot{\psi}^2 - \dot{\theta}^2\right)\cos\theta - \left(\ddot{\theta} + 2\dot{\phi}\dot{\psi}\cos\theta\right)\sin\theta\right]\left[\frac{I_{y_g} - I_{z_g}}{2}\sin 2\phi\right]$$

$$+ m_r\left[\ddot{\psi}R(\bar{x}_r\sin\theta - \bar{z}_r\cos\theta) - G\bar{y}_r\right] + \dot{\theta}\dot{\psi}\left[I_{y_r} + \left(I_{x_r} - I_{z_r}\right)\cos 2\theta - 2I_{xz_r}\sin 2\theta\right]$$

$$+ \ddot{\psi}\left[\frac{I_{x_r} - I_{z_r}}{2}\sin 2\theta + I_{xz_r}\cos 2\theta\right] + \ddot{\theta}\left[I_{xy_{r+}}\cos\theta + I_{yz_r}\sin\theta\right]$$

$$+ \left(\dot{\psi}^2 - \dot{\theta}^2\right)\left[I_{xy_r}\sin\theta - I_{yz_r}\cos\theta\right]$$

These forces and moments have the directions as shown in the diagram of pg. 1. In terms of the notation used in the structural dynamic study, we have:

$$\left.\begin{aligned} B_{U_0} &= F_Y(t) \\ A_{U_0} &= F_Z(t) \\ M_{R_U} &= -T_X(t) \\ M_{B_U} &= T_Y(t) \\ M_{A_U} &= T_Z(t) \end{aligned}\right\} \qquad \text{Ref. page 1, Chapter 1.}$$

4.3 Rigid Body Kinematic Loads Acting on The Gimbal Ring

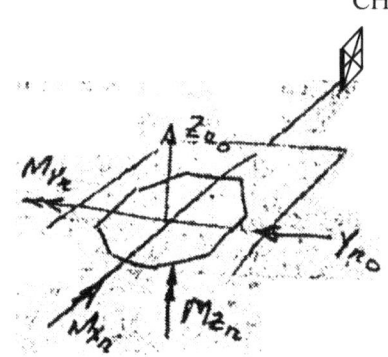

$$Y_{r_0} = m_r \left\{ \ddot{\psi} \left[R - \left(\overline{x}_r \cos\theta + \overline{z}_n \sin\theta \right) \right] + \dot{\psi}^2 \overline{y}_r \right.$$
$$\left. + 2\dot{\theta}\dot{\psi} \left(\overline{x}_r \sin\theta - \overline{z}_n \cos\theta \right) \right\} + m_g \left\{ \ddot{\psi} R \right\}$$

$$Z_{r_0} = -m_r \left\{ \dot{\psi}^2 \sin\theta \left[R - \left(\overline{x}_r \cos\theta + \overline{z}_n \sin\theta \right) \right] - \ddot{\psi}\overline{y}_r \sin\theta - \dot{\theta}^2 z_r - \ddot{\theta}\overline{x}_n + G\cos\theta \right\}$$
$$- m_g \left\{ \dot{\psi}^2 \sin\theta R + G\cos\theta \right\}$$

$$M_{x_r} = -m_r \left[\ddot{\psi} R I_r + \left(\dot{\psi}^2 R \sin\theta + G\cos\theta \right) \overline{y}_r \right] + \ddot{\psi} \left[I_{x_r} \sin\theta - I_{xz_r} \cos\theta \right]$$
$$+ \dot{\theta}\dot{\psi} \left[\left(I_{x_r} + I_{y_r} - I_{z_r} \right) \cos\theta - 2 I_{xz_r} \sin\theta \right] + \dot{\psi}^2 \cos\theta \left[I_{xy_r} \sin\theta - I_{yz_r} \cos\theta \right]$$
$$+ \ddot{\theta} I_{xy_r} + \dot{\theta}^2 I_{yz_r} + I_{x_g} \left[\ddot{\psi} \sin\theta + \dot{\theta}\dot{\psi} \cos\theta - \ddot{\phi} \right]$$
$$+ \dot{\theta}\dot{\psi} \cos\theta \left[\left(I_{y_g} - I_{z_g} \right) \right] \cos 2\phi + \left[\dot{\psi}^2 \cos^2\theta - \dot{\theta}^2 \left(\frac{I_{y_g} - I_{z_g}}{2} \right) \right] \sin 2\phi$$

$$M_{y_r} = m_r \left[\dot{\psi}^2 R \left(\overline{x}_r \sin\theta - \overline{z}_r \cos\theta \right) + G \left(\overline{x}_r \cos\theta + \overline{z}_r \sin\theta \right) \right] - \ddot{\theta} I_{y_r}$$
$$+ \dot{\psi}^2 \left[\frac{I_{x_r} - I_{z_r}}{2} \sin 2\theta + I_{xz_r} \cos 2\theta \right] - \ddot{\psi} \left[I_{xy_r} \sin\theta - I_{yz_r} \cos\theta \right]$$
$$I_{xy} \left[\left(\dot{\psi} \sin\theta - \dot{\phi} \right) \dot{\psi} \cos\theta \right] - \left[\ddot{\theta} + \dot{\psi}^2 \frac{\sin 2\theta}{2} \right] \left(\frac{I_{y_g} - I_{z_g}}{2} \right)$$
$$- \left[\ddot{\theta} + \left(2\dot{\phi} - \dot{\psi} \sin\theta \right) \dot{\psi} \cos\theta \right] \left(\frac{I_{y_g} - I_{z_g}}{2} \right) \cos 2\phi - \left[\ddot{\psi} \cos\theta - 2\dot{\theta}\dot{\phi} \right] \left(\frac{I_{y_g} - I_{z_g}}{2} \right) \sin 2\phi$$

$$M_{z_r} = m_r \left[\ddot{\psi} R \overline{x}_r + \left(\dot{\psi}^2 R \cos\theta - G\sin\theta \right) \overline{y}_r \right] - \ddot{\psi} \left[I_{z_r} \cos\theta + I_{xz_r} \sin\theta \right]$$
$$+ \dot{\theta}\dot{\psi} \left[\left(I_{y_r} + I_{z_r} - I_{x_r} \right) \sin\theta - 2 I_{xz_r} \cos\theta \right] + \dot{\psi}^2 \sin\theta \left[I_{xy_r} \sin\theta - I_{yz_r} \cos\theta \right]$$
$$- \dot{\theta}^2 I_{xy_r} + \ddot{\theta} I_{yz_r} - I_{x_g} \left[\dot{\theta} \left(\dot{\psi} \sin\theta - \dot{\phi} \right) \right] - \left[\dot{\psi} \cos\theta - 2\dot{\theta}\dot{\psi} \sin\theta \right] \left(\frac{I_{y_g} + I_{z_g}}{2} \right)$$
$$+ \left[\ddot{\psi} \cos\theta - 2\dot{\theta}\dot{\phi} \right] \left(\frac{I_{y_g} - I_{z_g}}{2} \right) \cos 2\phi - \left[\ddot{\theta} + \left(2\dot{\phi} - \dot{\psi} \sin\theta \right) \dot{\psi} \cos\theta \right] \left(\frac{I_{y_g} - I_{z_g}}{2} \right) \sin 2\phi$$

In terms of the load convention used in the dynamic study, we have:

$$\left.\begin{array}{l} Z_{r_0} = F_Z^*(t) \\[4pt] Y_{r_0} = -F_Y^*(t) \\[4pt] M_{X_r} = -T_X^*(t) \\[4pt] M_{Y_r} = -T_Y^*(t) \\[4pt] M_{Z_r} = T_Z^*(t) \end{array}\right\} \qquad \text{Ref. Page 1, Chapter 1}$$

* => Rotated through the angle θ from the arm axes.

4.4   Restraints on the System

Drives:

1st Axis:

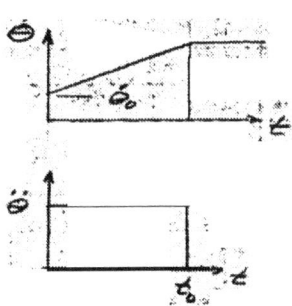

$$\dot{\theta}_{Max} = \frac{30 \times 2\pi}{60} = 3.14 \quad rad/\sec$$

$$\ddot{\theta}_{Max} = 1.4 \quad rad/\sec^2$$

Max duration: $t_0 = \dfrac{\dot{\theta}_{Max} - \dot{\theta}_0}{\ddot{\theta}(\leq 1.4)}$

$$\boxed{t_0 = \frac{3.14 - \dot{\theta}_0}{\ddot{\theta}(\leq 1.4)}}$$

$$\boxed{\ddot{\theta}}$$

For $\dot{\theta}_0 = 0, \quad \ddot{\theta} = 1.4$

$t_0 = 2.24 \quad \sec.$

2nd Axis: $\dot{\phi}_{Max} = 3.14 \quad rad/\sec$

$\ddot{\phi}_{Max} = 6.4 \quad rad/\sec$

Max duration: $t_0 = \dfrac{\dot{\phi}_{Max} - \dot{\phi}_0}{\ddot{\phi}(\leq 6.4)}$

$$\boxed{t_0 = \frac{3.14 - \dot{\phi}_0}{\ddot{\phi}(\leq 6.4)}}$$

$$\boxed{\ddot{\phi}}$$

Main Drive:

$\dot{\psi}_{Max} = 1.4 \times 2\pi/60 = 4.4 \quad rad/\sec$ $\ddot{\psi}_{Max} = 1.4 \quad rad/\sec$

$$\boxed{t_0 = \frac{4.4 - \dot{\psi}_0}{\ddot{\psi}(\leq 1.4)}}$$

$$\boxed{\ddot{\psi}}$$

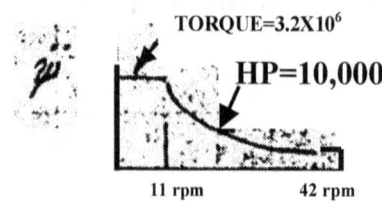

TORQUE=3.2X10^6

HP=10,000

11 rpm   42 rpm

But subject to restraint that $\ddot{\psi} = f(\dot{\psi})$

Brakes:

Braking torque 1st Axis = 1320 ft. lbs.
"   "   2nd Axis = 2200 ft. lbs.

Durations are variable.

4.5 Analysis for Transient Loads

If $\ddot{\psi} = \ddot{\theta} = \ddot{\phi} = 0$, and

$\quad \dot{\psi}_0 = \dot{\theta}_0 = \dot{\phi}_0 = $ arbitrary,

It can be seen from the rigid body kinematic loads that a stable configuration exists; i.e., rigid body motion of the system occurs without structural vibrations, since no periodic or transient forcing functions exist.

If either $\ddot{\psi}$, or $\ddot{\theta}$, $\ddot{\phi}$ is now imposed on the centrifuge, then structural vibrations will occur.

Let us consider first the case where a step input (rectangular pulse) in one of the angular accelerations is imposed.
    Call the acceleration "$\ddot{\omega}$".

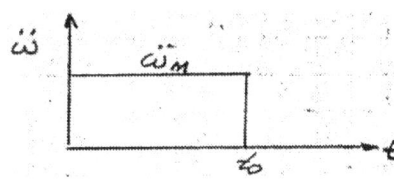

Then $\begin{cases} \ddot{\omega} = \ddot{\omega}_m, & 0 \leq t \leq t_0 \\ \ddot{\omega} = 0, & t < 0 \quad or \quad t > t_0 \end{cases}$

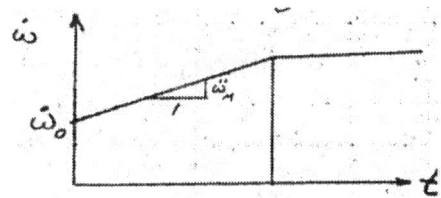

And $\begin{cases} \dot{\omega} = \ddot{\omega}_m t + \dot{\omega}_0 & , \; t \leq t_0 \\ \dot{\omega} = \ddot{\omega}_m t_0 + \dot{\omega}_0 & , \; t > t_0 \end{cases}$

Now, looking at the rigid body kinematic loads, we see that transient loads result from terms involving $\ddot{\omega}$ and $\dot{\omega}$. The former will have a step shape, while the latter will have the shape shown in the figure below; i.e., a ramp rise function.

$$F(\ddot{\omega})$$

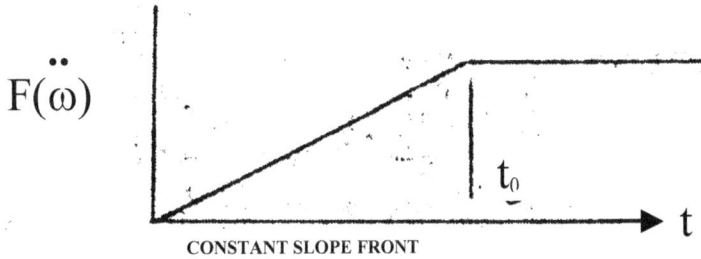

CONSTANT SLOPE FRONT

89

Referring to page 7, which delineates the limitations imposed on the systems performance, we observe that

$$\ddot{\psi} \leq 1.4 \quad rad/\sec^2, \qquad \ddot{\theta} \leq 1.4 \quad rad/\sec^2$$
$$\ddot{\phi} \leq 6.4 \quad rad/\sec^2$$

The most severe loading, from a structural dynamic response standpoint, will occur when the initial conditions are null (i.e., $\dot{\omega}_0 = 0$), since the response would involve the maximum static load term dependent on $\dot{\omega}$. However, since the max angular accelerations are small, the time of application of the ramp, t_0, will be relatively long compared to the fundamental period of the structure. For example, since:

$$t_{0_{Min}} = \frac{\dot{\omega}_{Max} - \dot{\omega}_0}{\ddot{\omega}_{Max}} = \frac{\dot{\omega}_{Max}}{\ddot{\omega}_{Max}}$$

and $\quad \dot{\psi}_{Max} = 4.4 \quad rad/\sec$, we have for the main axis:

$$t_{0_{Min}} = \frac{\dot{\psi}_{Max}}{\ddot{\psi}} = \frac{4.4}{1.4} = 3.14 \sec$$

since $f_1 \, \vartheta \, 2.5 \, cps, \quad T_1 \, \vartheta \, \frac{1}{2.5} \cong 0.40 \sec$

Hence $\left. \dfrac{t_0}{T_1} \right)_{Min} \; >> 1, \quad$ and static response of the system will result similarly, for the 2nd axis:

$$t_{0_{Min}} = \frac{\dot{\phi}_{Max}}{\ddot{\phi}_{Max}} = \frac{3.14}{6.4} = 0.49 \sec$$

and $\quad \left. \dfrac{t_0}{T_1} \right)_{Min} \geq \dfrac{.49}{.40} \approx 1.2$

Now, referring to Reference 4, "Shock and Vibration Handbook", Figure 8.13, a spectrum of maximax response resulting from step excitation functions with variously shaped rise fronts, we see that the dynamic response factor for the "constant slope front" is:

$$\gamma \leq 1.23$$

As a result, we conclude that the terms involving the ramp rise-sustained velocities $\dot{\phi}$ have, at the maximum, dynamic response factors of 1.2, while those involving $\dot{\phi}$ and $\dot{\psi}$ have factors of unity associated with them. Since the terms involving $\dot{\phi}$ in the load equations are not of primary significance (instead the acceleration terms are more important), we will neglect all load terms involving the velocities and assume static response to such terms.

Now, let us consider angular accelerations applied with a different time dependency. For shapes such as the ramp and half sine, shown below:

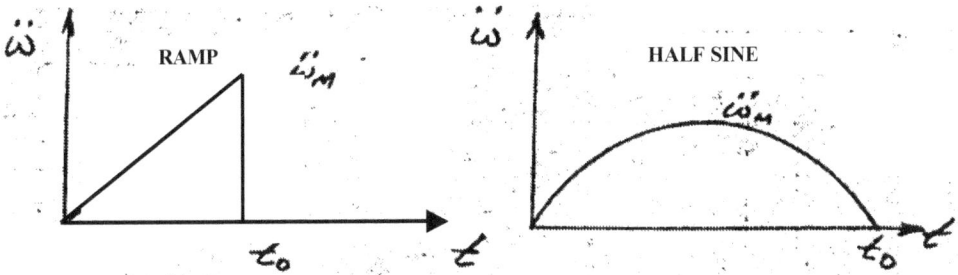

The velocity changes will be as shown below:

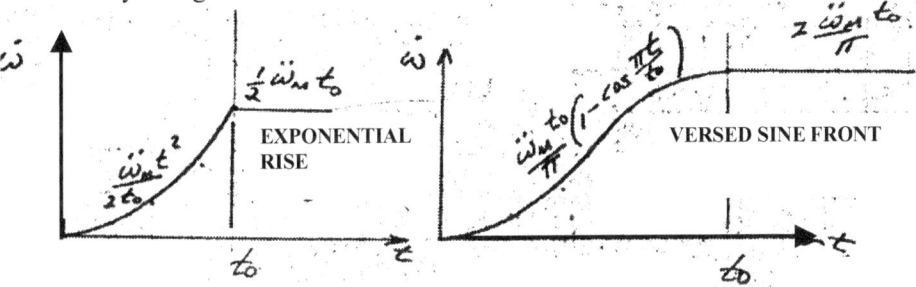

The dynamic response factor corresponding to the velocity diagram for the ramp acceleration can be obtained from Figure 8.14 of Reference 4, for an exponential rise with $+\pi \le a \le \dfrac{+\pi}{2}$.

We find that, for $\dfrac{t_0}{T_1} \ge 1.2$, the dynamic response T_1 factor is:

$$\gamma \le 1.30$$

Similarly, for the "versed sine front" velocity shape we have, from Figure 8.13:

$$\gamma \le 1.04$$

Hence, we reach the same conclusion for these diagrams as for the "constant slope front", that is, that structural dynamics need be investigated only for the terms in the rigid body loads which involve the angular accelerations. All other rigid body loadings may be considered as static applied loads. Proceeding on this basis, the transient loads on the arm and on the gimbal are as follows:

Transient Loads acting on the Main Arm:

$$F_Y(t) = B_{U_0} = -m_g \ddot{\psi} R - m_r \left\{ \ddot{\psi} \left[R - (\bar{x}_n \cos\theta + \bar{z}_n \sin\theta) \right] \right\}$$

$$T_Y(t) = M_{B_U} = \ddot{\theta} \left[\left(\frac{I_{y_g} + I_{z_g}}{2} \right) + \left(\frac{I_{y_g} - I_{z_g}}{2} \right) \cos 2\phi + I_{y_r} \right]$$

$$+ \ddot{\psi} \left[\cos\theta \left(\frac{I_{y_g} - I_{z_g}}{2} \right) \sin 2\phi + I_{xy_r} \sin\theta - I_{yz_r} \cos\theta \right]$$

$$F_Z(t) = A_{U_0} = +m_r \ddot{\theta}(\bar{x}_r \cos\theta + \bar{z}_r \sin\theta)$$

$$T_Z(t) = M_{A_U} = -I_{x_g} \ddot{\psi} \sin^2\theta - \left(\frac{I_{y_g} + I_{z_g}}{2} \right) \ddot{\psi} \cos^2\theta + \left(\frac{I_{y_g} - I_{z_g}}{2} \right) \ddot{\psi} \cos^2\theta$$

$$+ \cos 2\phi - \left(\frac{I_{y_g} - I_{z_g}}{2} \right) \ddot{\theta} \cos\theta \sin 2\phi + m_r \ddot{\psi} R(\bar{x}_r \cos\theta + \bar{z}_r \sin\theta)$$

$$- \ddot{\psi} \left[\left(\frac{I_{x_r} + I_{z_r}}{2} \right) - \left(\frac{I_{x_r} - I_{z_r}}{2} \right) \cos 2\theta + I_{xz_r} \sin 2\theta \right] - \ddot{\theta} \left[I_{xy_r} \sin\theta - I_{yz_r} \cos\theta \right]$$

$$= \ddot{\psi} \left\{ -I_{x_g} \sin^2\theta + \cos^2\theta \left[\cos 2\phi \left(\frac{I_{y_g} - I_{z_g}}{2} \right) - \left(\frac{I_{y_g} + I_{z_g}}{2} \right) \right] \right\}$$

$$+ m_r R(\bar{x}_r \cos\theta + \bar{z}_r \sin\theta) - \left[\left(\frac{I_{x_r} + I_{z_r}}{2} \right) - \left(\frac{I_{x_r} - I_{z_r}}{2} \right) \cos 2\theta + I_{xz_r} \sin 2\theta \right]$$

$$- \ddot{\theta} \left\{ \left(\frac{I_{y_g} - I_{z_g}}{2} \right) \cos\theta \sin 2\phi + I_{xy_r} \sin\theta - I_{yz_r} \cos\theta \right\}$$

$$T_X(t) = -M_{R_U} = -I_{x_g} \ddot{\psi} \sin\theta \cos\theta + \ddot{\psi} \sin\theta \cos\theta \left(\frac{I_{y_g} + I_{z_g}}{2} \right)$$

$$- \ddot{\psi} \sin\theta \cos\theta \left(\frac{I_{y_g} - I_{z_g}}{2} \right) \cos 2\phi + \ddot{\theta} \sin\theta \sin 2\phi \left(\frac{I_{y_g} - I_{z_g}}{2} \right)$$

$$- m_r \left[\ddot{\psi} R(\bar{x}_r \sin\theta - \bar{z}_r \cos\theta) \right] - \ddot{\psi} \left[\left(\frac{I_{x_r} - I_{z_r}}{2} \right) \sin 2\theta + I_{xz_r} \cos 2\theta \right]$$

$$- \ddot{\theta} \left[I_{xy_r} \cos\theta + I_{yz_r} \sin\theta \right] + \ddot{\phi} \cos\theta \, I_{x_g}$$

$$= \ddot{\psi} \left\{ \frac{\sin 2\theta}{2} \left(\frac{I_{y_g} + I_{z_g}}{2} - I_{x_g} \right) - \frac{\sin 2\theta}{2} \cos 2\phi \left(\frac{I_{y_g} - I_{z_g}}{2} \right) - m_r R(\bar{x}_r \sin\theta - \bar{z}_r \cos\theta) \right.$$

$$\left. - \left[\left(\frac{I_{x_r} - I_{z_r}}{2} \right) \sin 2\theta + I_{xz_r} \cos 2\theta \right] \right\} + \ddot{\theta} \left\{ \sin\theta \sin 2\phi \left(\frac{I_{y_g} - I_{z_g}}{2} \right) - I_{xy_r} \cos\theta - I_{yz_r} \sin\theta \right\}$$

Transient Loads Acting on the Gimbal Ring

$$F_Y \quad (t) = -Y_{r_0} = -m_r \ddot{\psi}\left[R - (\overline{x}_r \cos\theta + \overline{z}_r \sin\theta)\right] - m_g \ddot{\psi}R$$

$$T_Y(t) = -M_{Y_r} = \ddot{\theta}I_{y_r} + \ddot{\psi}\left[I_{xy_r} \sin\theta - I_{yz_r} \cos\theta\right]$$

$$\ddot{\theta}\left\{\left(\frac{I_{y_g} + I_{z_g}}{2}\right) + \left(\frac{I_{y_g} - I_{z_g}}{2}\right)\cos 2\phi\right\} + \ddot{\psi}\cos\theta\sin 2\phi\left(\frac{I_{y_g} - I_{z_g}}{2}\right)$$

$$= \ddot{\psi}\left\{I_{xy_r}\sin\theta - I_{yz_r}\cos\theta + \cos\theta\sin 2\phi\left(\frac{I_{y_g} - I_{z_g}}{2}\right)\right\}$$

$$+ \ddot{\theta}\left\{I_{y_r} + \left(\frac{I_{y_g} + I_{z_g}}{2}\right) + \left(\frac{I_{y_g} - I_{z_g}}{2}\right)\cos 2\phi\right\}$$

$$F_Z(t) = Z_{r_0} = +m_r\left\{\ddot{\psi}\overline{y}_r \sin\theta + \ddot{\theta}\overline{x}_r\right\}$$

$$T_Z(t) = M_{Z_r} = m_r\ddot{\psi}R\overline{x}_r - \ddot{\psi}\left[I_{z_r}\cos\theta + I_{xz_r}\sin\theta\right] + \ddot{\theta}I_{yz_r}$$

$$+ \ddot{\psi}\cos\theta\cos 2\phi\left(\frac{I_{y_g} - I_{z_g}}{2}\right) - \ddot{\theta}\sin 2\phi\left(\frac{I_{y_g} - I_{z_g}}{2}\right)$$

$$= \ddot{\psi}\left\{m_r R\overline{x}_r - \left(I_{z_r}\cos\theta + I_{xz_r}\sin\theta\right) + \cos\theta\cos 2\phi\left(\frac{I_{y_g} - I_{z_g}}{2}\right)\right\}$$

$$+ \ddot{\theta}\left\{I_{yz_r} - \sin 2\phi\left(\frac{I_{y_g} - I_{z_g}}{2}\right)\right\}$$

$$T_X(t) = -M_{X_r} = m_r\left[\ddot{\psi}R\overline{z}_r\right] - \ddot{\psi}\left[I_{x_r}c\sin\theta + I_{xz_r}\cos\theta\right] + \ddot{\theta}\,I_{yz_r}$$

$$- I_{x_g}\ddot{\psi}\sin\theta + I_{x_g}\ddot{\phi}$$

$$= \ddot{\psi}\left[m_r R\overline{z}_r - \left(I_{x_r}\sin\theta + I_{xz_r}\cos\theta\right) - I_{x_g}\sin\theta\right]$$

$$- \ddot{\theta}\left(I_{xy_r}\right) + \ddot{\phi}\left(I_{x_g}\right)$$

4.6 Summary- Transient Loads Acting on the Arm

Axis	Load	Due to $\ddot{\psi}$	$\ddot{\theta}$	$\ddot{\phi}$
Z	$F_Z(t)$	0	$m_r\left(\bar{x}_r\cos\theta + \bar{z}_r\sin\theta\right)$	0
	$T_Y(t)$	$\cos\theta\sin2\phi\left(\dfrac{I_{y_g}-I_{z_g}}{2}\right)+I_{xy_r}\sin\theta$ $-I_{yz_r}\cos\theta$	$\dfrac{I_{y_g}+I_{z_g}}{2}+$ $\dfrac{I_{y_g}-I_{z_g}}{2}\cos2\phi+I_{y_r}$	0
Y	$F_Y(t)$	$-R\left(m_g+m_r\right)+m_r\left(\bar{x}_r\cos\theta+\bar{z}_r\sin\theta\right)$	0	0
	$T_Z(t)$	$-I_{x_g}\sin^2\theta+\cos^2\theta\left[\cos2\phi\left(\dfrac{I_{y_g}-I_{z_g}}{2}\right)\right]$ $-\left(\dfrac{I_{y_g}+I_{z_g}}{2}\right)+m_rR\left(\bar{x}_r\cos\theta+\bar{z}_r\sin\theta\right)$ $-\left[\left(\dfrac{I_{x_r}+I_{z_r}}{2}\right)-\left(\dfrac{I_{x_r}-I_{z_r}}{2}\right)\cos2\theta+I_{xz_r}\sin2\theta\right]$	$I_{yz_r}\cos\theta-I_{xy_r}\sin\theta$ $-\left(\dfrac{I_{y_g}-I_{z_g}}{2}\right)\cos\theta\sin2\phi$	0
X	$T_X(t)$	$\dfrac{\sin2\theta}{2}\left(\dfrac{I_{y_g}+I_{z_g}}{2}-I_{x_g}\right)$ $-\dfrac{\sin2\theta}{2}\cos2\phi\left(\dfrac{I_{y_g}-I_{z_g}}{2}\right)$ $-m_rR\left(\bar{x}_r\sin\theta-\bar{z}_r\cos\theta\right)$ $-\left[\left(\dfrac{I_{x_r}-I_{z_r}}{2}\right)\sin2\theta+I_{xz_r}\cos2\theta\right]$	$\sin\theta\sin2\phi\left(\dfrac{I_{y_g}-I_{z_g}}{2}\right)$ $-I_{xy_r}\cos\theta-I_{yz_r}\sin\theta$	$\cos\theta\, I_{x_g}$

TABLE VIII

This is the final tabulation of transient loads acting on the system, for use in the analysis of Chapters 1 and 2.

4.7   Summary – Transient Loads Acting on the Gimbal Ring

Axis	Load	Due to $\ddot{\psi}$	Due to $\ddot{\theta}$	Due to $\ddot{\phi}$
Z	$F_Z(t)$	$m_r \bar{y}_r \sin\theta$	$m_r \bar{x}_r$	0
	$T_Y(t)$	$I_{xy_r}\sin\theta - I_{yz_r}\cos\theta$ $+\cos\theta\sin 2\phi\left(\dfrac{I_{y_g}-I_{z_g}}{2}\right)$	$I_{y_r}+\left(\dfrac{I_{y_g}+I_{z_g}}{2}\right)$ $+\left(\dfrac{I_{y_g}-I_{z_g}}{2}\right)\cos 2\phi$	0
Y	$F_Y(t)$	$-R(m_r+m_g)+m_r(\bar{x}_r\cos\theta+\bar{z}_r\sin\theta)$	0	0
	$T_Z(t)$	$m_r R\bar{x}_r-(I_{z_r}\cos\theta+I_{xz_r}\sin\theta)$ $+\cos\theta\cos 2\phi\left(\dfrac{I_{y_g}-I_{z_g}}{2}\right)$	$I_{yz_r}-\sin 2\phi\left(\dfrac{I_{y_g}-I_{z_g}}{2}\right)$	0
X	$T_X(t)$	$m_r R\bar{z}_r-(I_{x_r}\sin\theta+I_{xz_r}\cos\theta)$ $-I_{x_g}\sin\theta$	$-I_{xy_r}$	I_{x_g}

TABLE IX

Note: These loads act along gimbal coordinates, which are rotated by θ from arm coordinates. These loads are included for reference purposes only, or for use in a more detailed analysis of system response.

4.8 Study of Harmonic Forcing Functions

Examining the load equations summarized in section VI, it can be seen that harmonic forcing functions result from steady application of angular velocities $\dot{\theta}$ or $\dot{\phi}$.

We note from section IV that:

$$\dot{\theta}_{Max} = \pi \quad rad\,/\sec, \qquad \dot{\phi}_{Max} = \pi \quad rad\,/\sec$$

Hence the maximum harmonic frequency is

$$f_{Max} = \frac{\omega_{Max}}{2\pi} = \frac{\pi}{2\pi} = \frac{1}{2}\,cps$$

Since the lowest frequency of η system is

$$f_N \,\vartheta\, 2.50 \ cps\ ,$$

$$\frac{f_{Max}}{f_N} \,\vartheta\, \frac{1\!\!\!/_2}{2.50} = 0.20$$

Now, the steady state response of a simple oscillator to a harmonic forcing function with this ratio of implied to natural frequency will be a static response, or very close to that (i.e., magnification factor $\cong 1.0$).

Hence, harmonic forcing functions may be neglected in the structural dynamic analysis and analyzed as static applied loadings.

Chapter 5 Conclusions

A preliminary study of the structural dynamic response of the NASDA-Houston MSC centrifuge to transient loads induced by rotations of the gondola during rotation of the centrifuge arm has been completed.

The results of the study are summarized in the "Tables of Dynamic Response Factors" for the arm for lateral, vertical and torsional modes for three types of typical impulsive, generalized forcing functions. They are:

1. Square step (better described as a rectangular pulse).

2. Ramp.

3. Half Sine.

In part C of the study, the rigid body kinematic equations have been analyzed so that the generalized loading functions employed in the analysis may be interpreted in terms of the motion parameters of the kinematic analysis.

Future work would include the derivation of detail loads acting on the components based on the accelerations derived in this preliminary study and the completion of a structural analysis of the various components present in the assembly.

Figure 7.1 in Appendix 7 is a photograph of the MSC centrifuge in its deployed but non-rotating configuration.

Bibliography

R.W.Clough & J.Penzien, *Dynamics of Structures,* McGraw-Hill, New York, 1975.

J.H.Ginsberg, *Advanced Engineering Dynamics,* 2nd Ed., Cambridge University Press, New York, 1995.

C.M.Harris & C.E.Crede, *Shock and Vibration Handbook,* John Wiley & Sons, New York, 1986.

C.Lanczos, *The Variational Principles of Mechanics,* 4th Ed., Dover, New York, 1986.

J.N.Reddy, *Energy Principles and Variational Methods,* 2nd Ed., John Wiley & Sons, New York, 2002.

R.A.Shabana, *Dynamics of Multi-Body Systems,* John Wiley & Sons, New York, 1986.

K.J.Walden & G.L.Kinzel, *Kinematics, Dynamics & Design of Mechanisms,* John Wiley & Sons, New York, 2003.

D.Wells, *Lagrangian Dynamics,* John Wiley & Sons, New York, 1967.

Appendix 1

Derivation of the Y Axis Equations
of Motion for the Arm Model

A.1.1 Derivation of Motion Equations

Y axis

MODEL IN Y PLANE

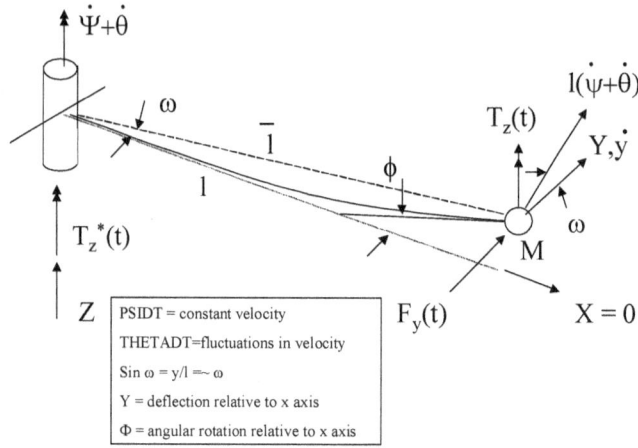

FIGURE A.1

Velocity of mass 'm': $\qquad \bar{v} = \ell(\dot{\psi} + \dot{\theta}) \quad + > y$

$\therefore \quad v^2 = \ell^2(\dot{\psi} + \dot{\theta}) + \dot{y}^2 + 2\ell\dot{y}(\dot{\psi} + \dot{\theta})\cos\omega$

Now $\quad \cos\omega = 1 - \dfrac{y^2}{2\ell^2} \qquad$ for $\omega \ll \pi/2$

Thus, the kinetic energy of the system is given by:

$$T = \frac{1}{2}I_{Z_R}(\dot{\psi} + \dot{\theta})^2 + \frac{m}{2}\left\{\ell^2(\dot{\psi} + \dot{\theta})^2 + \dot{y}^2 + 2\ell\dot{y}(\dot{\psi} + \dot{\theta})\left(1 - \frac{y^2}{2\ell^2}\right)\right\} + \frac{1}{2}I_{Z_M}(\dot{\phi} + \dot{\psi} + \dot{\theta})^2 \quad (1)$$

The potential energy of the system consists of the strain energy due to the arm displacement:

$$V = \frac{1}{2}k_{11}y^2 + \frac{1}{2}k_{22}\phi^2 - k_{12}y\phi \tag{2}$$

Where k_1 and k_2 are spring constants defined as follows:

$$\delta y = \alpha_{11}F_y + \alpha_{15}T_z$$
$$\theta_z = \alpha_{51}F_y + \alpha_{55}T_z$$

$\alpha_{ij} =>$ influence coefficients (see appendix 2)

$$F_y = \frac{\alpha_{55}\delta y - \alpha_{15}\theta_z}{\alpha_{11}\alpha_{55} - \alpha_{15}\alpha_{51}} = k_{11}\delta y - k_{12}\theta_z$$

$$T_z = \frac{\alpha_{11}\theta_z - \alpha_{51}\delta y}{\alpha_{11}\alpha_{55} - \alpha_{15}\alpha_{51}} = -k_{21}\delta y + k_{22}\theta_z$$

$$\text{and } k_{11} = \alpha_{55}\big/\Delta, \quad k_{12} = \alpha_{15}\big/\Delta$$

$$k_{21} = k_{12}, \quad k_{22} = \alpha_{11}\big/\Delta$$

$$\Delta = \alpha_{11}\alpha_{55} - \alpha_{15}\alpha_{51}$$

since $y \equiv \delta y$, $\phi \equiv \theta_z$, Eq. (2) follows.

Lagrange's equations for the system are given as:

$$\frac{d}{dt}\frac{\partial T}{\partial \dot{q}_s} - \frac{\partial T}{\partial q_s} + \frac{\partial V}{\partial q_s} = Q_s \quad ; s= 1, 2, 3$$

Where $q_1 = f_1(t) + s_1$, $q_2 = f_2(t) + s_2$, $q_3 = f_3(t) + s_3$

And for steady motion of the system

$$q_1 = f_1(t) = \dot{\psi}t, \quad q_2 = f_2(t) = 0, \quad q_3 = f_3(t) = 0$$

And the disturbances are

$$s_1 = \theta, \quad s_2 = y, \quad s_3 = \phi$$

Using these quantities as new coordinates, we have the Lagrangian equations for the disturbed motion as:

$$\frac{d}{dt}\frac{\partial T}{\partial \dot{s}_i} - \frac{\partial T}{\partial s_i} + \frac{\partial V}{\partial s_i} = \Delta Q_s \tag{3}$$

Differentiating:

$$\left\{ \begin{array}{l} \dfrac{\partial T}{\partial \dot{s}_1} = \dfrac{\partial T}{\partial \dot{\theta}} = I_{Z_R}\left(\dot{\psi} + \dot{\theta}\right) + m\ell^2(\dot{\psi} + \dot{\theta}) + m\ell\dot{y}\left(1 - \dfrac{y^2}{2\ell^2}\right) + I_{Z_M}(\dot{\phi} + \dot{\psi} + \dot{\theta}) \\[2mm] \dfrac{\partial T}{\partial s_1} = \dfrac{\partial T}{\partial \theta} = 0 \\[2mm] \dfrac{\partial V}{\partial s_1} = \dfrac{\partial V}{\partial \theta} = 0 \end{array} \right.$$

$$\begin{cases} \dfrac{\partial T}{\partial \dot{s}_2} = \dfrac{\partial T}{\partial \dot{y}} = m\dot{y} + m\ell(\dot{\psi} + \theta)\left(1 - \dfrac{y^2}{2\ell^2}\right) \\ \qquad \dfrac{\partial T}{\partial s_2} = \dfrac{\partial T}{\partial y} = \dfrac{-m}{\ell}\,\dot{y}(\dot{\psi} + \dot{\theta})y \\ \qquad \dfrac{\partial V}{\partial s_2} = \dfrac{\partial V}{\partial y} = k_{11}y - k_{12}\phi \end{cases}$$

$$\begin{cases} \dfrac{\partial T}{\partial \dot{s}_3} = \dfrac{\partial T}{\partial \dot{\phi}} = I_{Z_M}\left(\dot{\phi} + \dot{\psi} + \dot{\theta}\right) \\ \dfrac{\partial T}{\partial s_3} = \dfrac{\partial T}{\partial \phi} = 0 \\ \dfrac{\partial V}{\partial s_3} = \dfrac{\partial V}{\partial \phi} = k_{22}\phi - k_{12}y \end{cases}$$

Substituting into the Lagrangian Eqs. (3)

$$\begin{cases} \dfrac{d}{dt}\left\{ I_{Z_R}(\dot{\psi} + \dot{\theta}) + m\ell^2(\dot{\psi} + \dot{\theta}) + m\ell\dot{y}\left(1 - \dfrac{y^2}{2\ell^2}\right) + I_{Z_M}(\dot{\phi} + \dot{\psi} + \dot{\theta})\right\} = T_Z^*(t) + \ell F_Y(t) + T_Z(t) \\ \dfrac{d}{dt}\left\{ m\dot{y} + m\ell(\dot{\psi} + \dot{\theta})\left(1 - \dfrac{y^2}{2\ell^2}\right)\right\} + \dfrac{m\dot{y}y}{\ell}(\dot{\psi} + \dot{\theta}) + k_{11}y - k_{12}\phi = F_Y(t) \\ \dfrac{d}{dt}\left\{ I_{Z_M}(\dot{\phi} + \dot{\psi} + \dot{\theta})\right\} + k_{22}\phi - k_{12}y = T_Z(t) \end{cases}$$

$$(4)$$

Performing the time differentiation:

$$\begin{cases} I_{Z_R}\ddot{\theta} + m\ell^2\ddot{\theta} + m\ell\ddot{y}\left(1 - \dfrac{y^2}{2\ell^2}\right) - \dfrac{m}{\ell}(\dot{y})^2 y + I_{Z_M}(\ddot{\phi} + \ddot{\theta}) = T_Z^*(t) + \ell F_Y(t) + T_Z(t) \\ m\ddot{y} + \ell\ddot{\theta}m\left(1 - \dfrac{y^2}{2\ell^2}\right) - \dfrac{2m}{\ell}\dot{y}y(\dot{\psi} + \dot{\theta}) + \dfrac{m\dot{y}y}{\ell}(\dot{\psi} + \dot{\theta}) + k_{11}y - k_{12}\phi = F_Y(t) \\ I_{Z_M}(\ddot{\phi} + \ddot{\theta}) + k_{22}\phi - k_{12}y = T_Z(t) \end{cases} \qquad (5)$$

These equations may be simplified by omitting powers or products of the small quantities $\theta, \phi, y, \dot{\theta}, \dot{\phi}, \dot{y},$ to obtain the reduced set:[*]

---

[*] Star superscript denote changes from steady state conditions.

$$\begin{cases} (I_{Z_R} + I_{Z_M} + m\ell^2)\ddot{\theta} + m\ell\ddot{y} + I_{Z_M}\ddot{\phi} = T_Z^*(t) + \ell F_Y(t) + T_Z(t) \\ m\ddot{y} + m\ell\ddot{\theta} + k_{11}y - k_{12}\phi = F_Y(t) \\ I_{Z_M}(\ddot{\theta} + \ddot{\phi}) + k_{22}\phi - k_{12}y = T_Z(t) \end{cases} \qquad (6)$$

letting $\begin{cases} y' = y + \ell\theta \\ \phi' = \phi + \theta \end{cases}$

we obtain, finally:

$$\begin{cases} I_{Z_R}\ddot{\theta} + m\ell\ddot{y}' + I_{Z_M}\ddot{\phi}' = T_Z^*(t) + \ell F_Y(t) + T_Z(t) \\ m\ddot{y}' + k_{11}(y' - \ell\theta) - k_{12}(\phi' - \theta) = F_Y(t) \\ I_{Z_M}\ddot{\phi}' + k_{22}(\phi' - \theta) - k_{12}(y' - \ell\theta) = T_Z(t) \end{cases} \qquad (7)$$

Equations (7) can be inverted to obtain them in terms of the influence coefficients. Thus,

$$k_{11}(y) - k_{12}(\phi) = F_Y(t) - m\ddot{y}'$$
$$-k_{12}(y) + k_{22}(\phi) = T_Z(t) - I_{Z_M}\ddot{\phi}'$$

or,

$$y' - \ell\theta = \begin{vmatrix} F_Y(t) - m\ddot{y}' & -k_{12} \\ T_Z(t) - I_{Z_M}\ddot{\phi}' & k_{22} \end{vmatrix} \Big/ \Delta$$

$$\phi' - \theta = \begin{vmatrix} k_{11} & F_Y(t) - m\ddot{y}' \\ -k_{12} & T_Z(t) - I_{Z_M}\ddot{\phi}' \end{vmatrix} \Big/ \Delta$$

where $\Delta = k_{11}k_{22} - k_{12}^2$

Now, $\dfrac{k_{22}}{\Delta} = \alpha_{11}, \quad \dfrac{k_{12}}{\Delta} = \alpha_{15}, \quad \dfrac{k_{11}}{\Delta} = \alpha_{55},$

Thus, we obtain the new set of equations:

$$\begin{cases} I_{Z_R}\ddot{\theta} + m\ell\ddot{y}' + I_{Z_M}\ddot{\phi}' = T_Z^*(t) + \ell F_Y(t) + T_Z(t) \\ y' - \ell\theta = \alpha_{11}F_Y(t) - \alpha_{11}m\ddot{y}' + \alpha_{15}T_Z(t) - \alpha_{15}I_{Z_M}\ddot{\phi}' \\ \phi' - \theta = \alpha_{15}F_Y(t) - \alpha_{15}m\ddot{y} + \alpha_{55}T_Z(t) - \alpha_{55}I_{Z_M}\ddot{\phi}' \end{cases} \qquad (8)$$

Changing to the model notation, we have finally:

103

$$\begin{cases} MR\ddot{y}_M + I_{Z_M}\ddot{\theta}_Z + I_{Z_R}\ddot{\psi}^* = T_Z^*(t) + RF_Y(t) + T_Z(t) \\ y_M - R\psi^* = -\alpha_{11}M\ddot{y}_M - \alpha_{15}I_{Z_M}\ddot{\theta}_Z + \alpha_{11}F_Y(t) + \alpha_{15}T_Z(t) \\ \theta_Z - \psi^* = -\alpha_{51}M\ddot{y}_M - \alpha_{55}I_{Z_M}\ddot{\theta}_Z + \alpha_{51}F_Y(t) + \alpha_{55}T_Z(t) \end{cases} \qquad (9)$$

since
$$\ddot{\theta} \Rightarrow \ddot{\psi}^*$$
$$\ddot{\phi}' \Rightarrow \ddot{\theta}_Z \qquad \ell \Rightarrow R$$
$$\ddot{y}' \Rightarrow \ddot{y}_M$$

Appendix 2

**Derivation of Flexibility
Matrix for the Arm**

Model

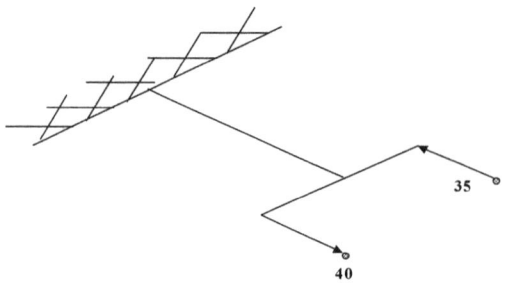

FIGURE A.2

Flexibility Derivation

From the computer results of deflections at nodes 35 & 40 due to loads applied at node 35, we may obtain the required influence coefficients.

<u>Case 2, computer print-out</u> $(F_Z = -1^{\#})$ All deflections shown are $\times 10^{-6}$

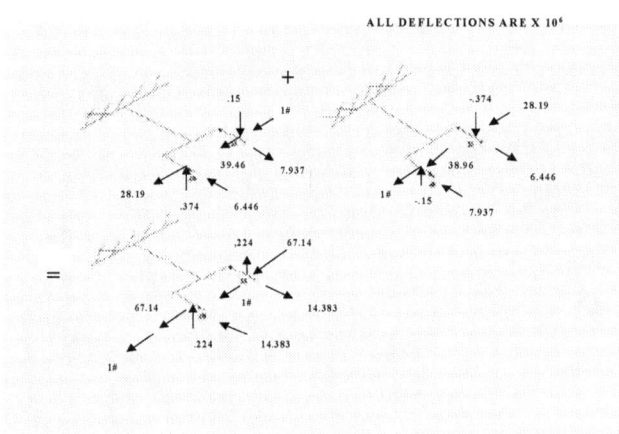

FIGURE A.3

<u>For $F_Z = -1^{\#}$ at center</u>

$$\delta x = 0$$

$$\delta y = -.224\big/_2 = -.112 \times 10^{-6}$$

$$\delta z = -38.57 \times 10^{-6}$$

$$\delta\theta_X = 0$$

$$\delta\theta_Y = 14.383\big/_{2 \times 108} \times 10^{-6}$$

$$\delta\theta_Z = 0$$

Case 1 $F_X = 1^{\#}$

ALL DEFLECTIONS ARE X 10^6

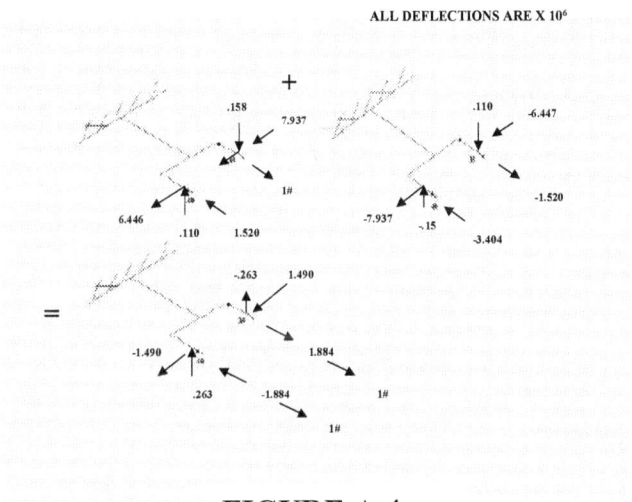

FIGURE A.4

For $F_X = 1^{\#}$ at center

$$\delta x = .942 \times 10^{-6}$$

$$\delta y = 0$$

$$\delta z = 0$$

$$\delta\theta_X = -.268\big/_{2 \times 108} \times 10^{-6}$$

$$\delta\theta_Y = 0$$

$$\delta\theta_Z = 0$$

<u>Case 3</u> $\left(F_Y = -1^{\#}\right)$

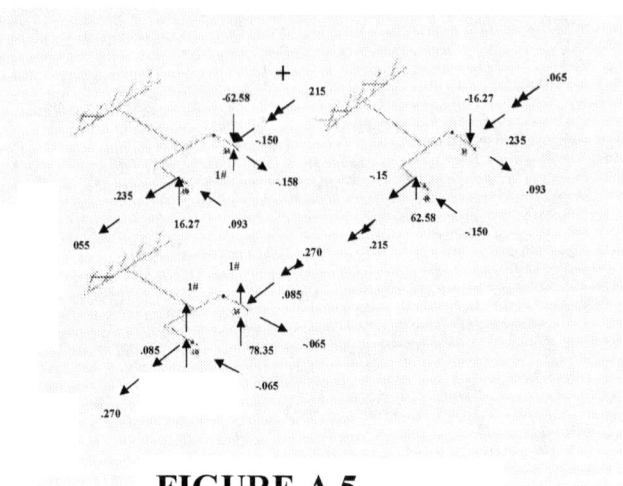

FIGURE A.5

<u>For $F_Y = -1^{\#}$ at center</u>

$$\delta x =$$
$$\delta y = -39.425 \times 10^{-6}$$
$$\delta z = -.0425 \times 10^{-6}$$
$$\delta\theta_X = 0$$
$$\delta\theta_Y = \frac{-.065}{2 \times 108} \times 10^{-6}$$
$$\delta\theta_Z = -.135 \times 10^{-6}$$

Case 4 $(M_Z = -1.0^\#)$

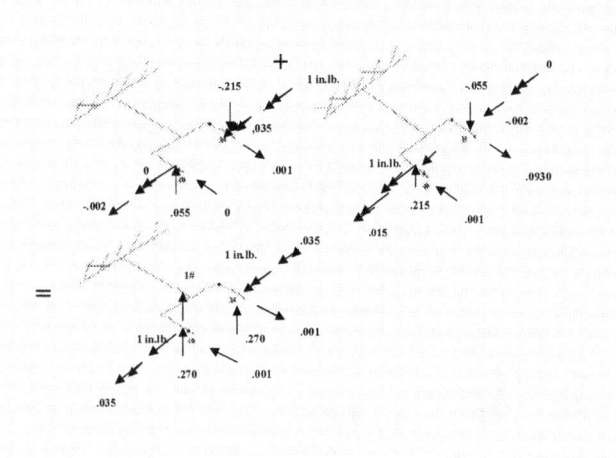

FIGURE A.6

<u>For $M_Z = -1_{in.}^\#$ at center</u>

$$\delta x = 0$$
$$\delta y = -.135 \times 10^{-6}$$
$$\delta z = 0$$
$$\delta\theta_X = 0$$
$$\delta\theta_Y \cong 0$$
$$\delta\theta_Z = -.0175 \times 10^{-6}$$

$\underline{M_x = 216\,_{in.}{}^{\#}}$ (use case 3: $F_y = -1^{\#}$ at 35, $+1^{\#}$ at 40)

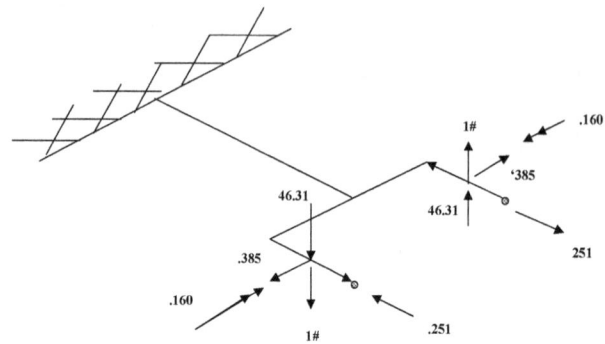

FIGURE A.7

$\underline{\text{For } M_x = 216\,_{in.}{}^{\#}}$

$$\delta x = -.251 \times 10^{-6}$$
$$\delta y = 0$$
$$\delta z = 0$$
$$\delta\theta_X = \frac{46.31}{108} \times 10^{-6}$$
$$\delta\theta_Y = 0$$
$$\delta\theta_Z = 0$$

$\underline{M_y = 216_{\text{in.}}{}^{\#}}$ (use case 1: $F_x = +1^{\#}$ at 35, $+1^{\#}$ at 40)

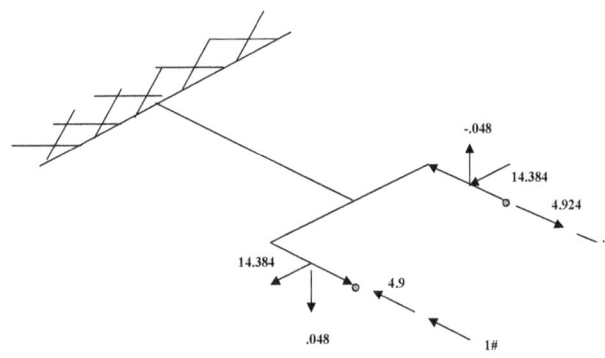

FIGURE A.8

$\underline{\text{For } M_x = 216_{\text{in.}}{}^{\#}}$

$$\delta x = 0$$
$$\delta y = .048 \times 10^{-6}$$
$$\delta z = -14.384 \times 10^{-6}$$
$$\delta \theta_X = 0$$
$$\delta \theta_Y = \frac{4.924}{108} \times 10^{-6}$$
$$\delta \theta_Z = 0$$

Summary of Influence Coefficients

$$[\alpha_{ij}][F]=[\delta] \tag{1}$$

Where $[\alpha_{ij}]$ = Flexibility matrix

$$[F]=\begin{bmatrix} F_X \\ F_Y \\ F_Z \\ M_X \\ M_Y \\ M_Z \end{bmatrix} ; \quad [\delta]=\begin{bmatrix} \delta_X \\ \delta_Y \\ \delta_Z \\ \theta_X \\ \theta_Y \\ \theta_Z \end{bmatrix} \tag{2}$$

Flexibility matrix

$$[\alpha_{ij}]^* = 10^{-6}\begin{bmatrix} .942 & 0 & 0 & \dfrac{-.259}{216} & 0 & 0 \\ 0 & 39.42 & .078 & 0 & \dfrac{.056}{216} & .135 \\ 0 & .078 & 33.57 & 0 & \dfrac{-14.384}{216} & 0 \\ \dfrac{-.259}{216} & 0 & 0 & \dfrac{46.31}{108\times216} & 0 & 0 \\ 0 & \dfrac{.056}{216} & \dfrac{-14.384}{216} & 0 & \dfrac{4.924}{108\times216} & 0 \\ 0 & .135 & 0 & 0 & 0 & .0175 \end{bmatrix} \tag{3}^*$$

Neglecting F_X, δ_X, the equations reduce to:

$$[F]=\begin{bmatrix} F_Y \\ F_Z \\ M_X \\ M_Y \\ M_Z \end{bmatrix} ; \quad [\delta]=\begin{bmatrix} \delta_Y \\ \delta_Z \\ \theta_X \\ \theta_Y \\ \theta_Z \end{bmatrix} \tag{4}$$

---

* Note: some slight changes in the coefficients have been made in order to insure reciprocity is satisfied.

$$[\alpha_{ij}] = 10^{-6} \begin{bmatrix} 39.42 & .078 & 0 & \dfrac{.056}{216} & .135 \\[2mm] .078 & 33.57 & 0 & \dfrac{-14.384}{216} & 0 \\[2mm] 0 & 0 & \dfrac{46.31}{108 \times 216} & 0 & 0 \\[2mm] \dfrac{.056}{216} & \dfrac{-14.984}{216} & 0 & \dfrac{4.924}{108 \times 216} & 0 \\[2mm] .135 & 0 & 0 & 0 & 0.0175 \end{bmatrix} \quad (5)$$

Now, we wish to make a coordinate transformation from the arm axes to the dynamic axes; i.e.,

ARM AXES   DYNAMICS AXES

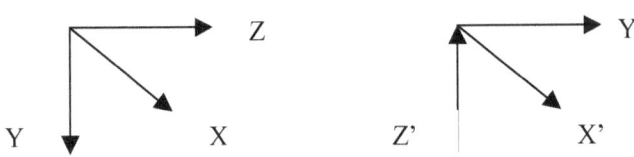

Let [A] => tensor transformation from x to x^1 coordinates, now,

$$[A][F] = [F^1]$$
$$[A][\delta] = [\delta^1]$$

Also, denoting $[\alpha_{ij}] => [\Im]$

We have, $[\Im][F] = [\delta]$

But, $[A^{-1}][F^1] = [F]$

Hence $[\Im][A^{-1}][F^1] = [\delta]$

And $[A][\Im][A^{-1}][F^1] = [\delta^1]$

Now, $[A] = \begin{bmatrix} 0 & 1 & 0 & 0 & 0 \\ -1 & 0 & 0 & 0 & 0 \\ 0 & 0 & 1 & 0 & 0 \\ 0 & 0 & 0 & 0 & 1 \\ 0 & 0 & 0 & -1 & 0 \end{bmatrix} = [A]^T \quad (6)$

 And

$$[A^{-1}] = \begin{bmatrix} 0 & -1 & 0 & 0 & 0 \\ 1 & 0 & 0 & 0 & 0 \\ 0 & 0 & 1 & 0 & 0 \\ 0 & 0 & 0 & 0 & -1 \\ 0 & 0 & 0 & 1 & 0 \end{bmatrix} = [A]^T \qquad (7)$$

Thus, denoting $\quad [A][\Im][A^{-1}] = [\Im^1]$, we have

$$[\Im^1] = 10^{-6} \begin{bmatrix} 33.57 & -.078 & 0 & 0 & \dfrac{14.384}{216} \\[2mm] -.078 & 39.42 & 0 & -.135 & \dfrac{.056}{216} \\[2mm] 0 & 0 & \dfrac{46.31}{108 \times 216} & 0 & 0 \\[2mm] 0 & -.135 & 0 & .0175 & 0 \\[2mm] \dfrac{14.384}{216} & \dfrac{.056}{216} & 0 & 0 & \dfrac{4.924}{108 \times 216} \end{bmatrix} \qquad (8)$$

This is the final flexibility matrix.

And $\quad [\Im^1][F^1] = [\delta^1]$ \hfill (9)

Where:

$$[\Im^1] = [\alpha_{ij}] ; \quad [F^1] = \begin{bmatrix} F_Y \\ F_Z \\ T_X \\ T_Y \\ T_Z \end{bmatrix} ; \quad [\delta^1] = \begin{bmatrix} \delta_Y \\ \delta_Z \\ \theta_X \\ \theta_Y \\ \theta_Z \end{bmatrix} \qquad (10)$$

Appendix 3

Investigation of Coupling Between Vertical, Lateral, and Torsional Modes

Employing the influence coefficients given in the flexibility matrix (5) of appendix 2, equations of free vibrational motion for the system can be written as follows:

$$[\delta] = [\alpha_{ij}][F] \qquad (1)$$

Now, $\quad [F] = -m[\ddot{X}] \qquad (2)$

Where $\qquad [\ddot{X}] = \begin{bmatrix} \delta_y \\ \delta_z \\ \ddot{\theta}_x \\ \ddot{\theta}_y \\ \ddot{\theta}_x \end{bmatrix} \qquad (3)$

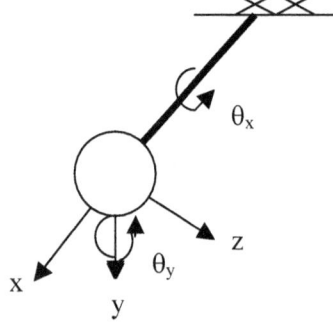

Using the Laplacian operator $s = \dfrac{d}{dt}$, and assuming initial conditions: $\quad [\delta] = 0 = [\dot{\delta}]$,

$$L\{[\ddot{X}]\} = s^2[u] \qquad (4)$$

Where $\qquad [u] = \begin{bmatrix} u_1 \\ u_2 \\ u_3 \\ u_4 \\ u_5 \end{bmatrix} = L\{[\delta]\} \qquad (5)$

Hence,

$$[u] = -m[\alpha_{ij}]s^2[u]$$

$$[I][u] + ms^2[\alpha_{ij}][u] = 0$$

or, $\qquad\qquad\qquad\qquad\qquad\qquad\qquad (6)$

$$\{[I] + ms^2[\alpha_{ij}]\}[u] = 0$$

The coupling between various modes will now be investigated.

A. Coupling between lateral and vertical modes, neglecting rotary inertia terms.

From Eqs. (5), we have:

$$\left(1+\alpha_{11}ms^2\right)u_1+\alpha_{12}ms^2u_2=0$$
$$\alpha_{21}ms^2u_1+\left(1+\alpha_{22}ms^2\right)u_2=0$$

The characteristic equation is:

$$s^4\left(\alpha_{11}\alpha_{22}-\alpha_{12}\alpha_{21}\right)m^2+s^2\left(\alpha_{11}+\alpha_{22}\right)m+1=0$$

A measure of the coupling present in the system is given by:

$$K=\frac{\alpha_{12}\alpha_{21}}{\alpha_{11}\alpha_{22}}$$

using the influence coefficients of appendix 2, (5):

$$K=\frac{\overline{.078}^2}{39\times33}\cong0$$

Thus, coupling effects are negligible, since the roots of the characteristic Eq. are independent of α_{12}, α_{21}.

B. Coupling between lateral and torsional modes.

$$\left(1+\alpha_{22}ms^2\right)u_2+\alpha_{23}ms^2u_3=0$$
$$\alpha_{32}ms^2u_2+\left(1+\alpha_{33}ms^2\right)u_3=0$$

since $\alpha_{23}=\alpha_{32}\equiv0$,   No Coupling Exists!

C. Coupling between vertical and torsional modes.

Since $\alpha_{13}=\alpha_{31}=0$,   No Coupling Exists.

Appendix 4

Computer Code and Print-Out
Transient Response

Problem definition

i)   For each of the specific values of t_{0_K} ; $k =1, 2, 3, ..., 15,$ calculate the values of the following functions versus time over the indicated range of the argument:

$$f_{11_K}(t) = 1 - \cos s_1 t \quad , \qquad 0 \le t < \frac{s_2}{s_1} t_{0_K} \qquad t3FE$$

$$f_{12_K}(t) = \cos s_1(t - t_{0_K}) - \cos s_1 t, \qquad t_{0_K} \le t \le \frac{s_2}{s_1} t_{Max} \qquad t2FE$$

$$f_{21_K}(t) = s_1 t - \sin s_1 t, \qquad 0 \le t < \frac{s_2}{s_1} t_{0_K}$$

$$f_{22_K}(t) = s_1 t \cos s_1(t - t_{0_K}) - \sin s_1 t - s_1(t - t_{0_K}) \cos s_1(t - t_{0_K}) + \sin s_1(t - t_{0_K}),$$

$$t_{0_K} \le t \le \frac{s_2}{s_1} t_{Max}$$

$$f_{31_K}(t) = \frac{2 t_{0_K}}{\pi^2 - s_1^2 t_{0_K}^2} \left[\pi \sin s_1 t - s_1 t_{0_K} \sin \frac{\pi t}{t_{0_K}} \right], \qquad 0 \le t \le \frac{s_2}{s_1} t_{0_K}$$

$$f_{32_K}(t) = \frac{2 t_{0_K} \pi}{\pi^2 - s_1^2 t_{0_K}^2} \left[\sin s_1 t (1 + \cos s_1 t_{0_K}) - \cos s_1 t \sin s_1 t_{0_K} \right] \qquad t_{0_K} \le t \le \frac{s_2}{s_1}$$

Where t_{Max}, Δt, s_1, and s_2 are constants, and

$$t_{0_K} = \frac{0.6283 K}{s_1}; \qquad k=1, 2, 3, ..., 15$$

Let the functions be denoted by:

ii)   FJLK(t) = $f_{jlk}(t)$, where

iii)

$$j = 1, 2, 3 \text{ denotes types of loadings}$$
$$l = 1, 2 \quad \text{denotes functional range}$$
$$k = 1, 2, ..., 15$$

We note that:

$$f_{1lk}(t) = F_{1_K}(t) \qquad \text{Eq. (83), text}$$
$$f_{2lk}(t) = G_{1_K}(t) \qquad \text{Eq. (84), text}$$
$$f_{3lk}(t) = H_{1_K}(t) \qquad \text{Eq. (85), text}$$

iv)   For each of the specific values of t_{0_K}, calculate the following values versus time, from t = 0, in increments of Δt :

For $0 \le t < t_{0_K}$:

$$YIJK = AJ1 \times \Gamma I1 \times FJ1K(t) + AJ2 \times \Gamma I2 \times FJ1\frac{s_2}{s_1} K\left(\frac{s_2}{s_1} t\right)$$

$$\theta IJK = AJ1 \times LI1 \times FJ1K(t) + AJ2 \times LI2 \times FJ1\frac{s_2}{s_1} K\left(\frac{s_2}{s_1} t\right)$$

For $t_{0_K} \le t < t_{Max}$: $\qquad\qquad\qquad\qquad$ Range $L\Big\{_2^1$

$$YIJK = AJ1 \times \Gamma I1 \times FJ2K(t) + AJ2 \times \Gamma I2 \times FJ2\frac{s_2}{s_1} K\left(\frac{s_2}{s_1} t\right)$$

$$\theta IJK = AJ1 \times LI1 \times FJ2K(t) + AJ2 \times LI2 \times FJ2\frac{s_2}{s_1} K\left(\frac{s_2}{s_1} t\right)$$

where:

$$A_{11} = 1.0; \quad A_{21} = \frac{1}{s_1 t_{0_K}}; \quad A_{31} = \frac{s_1}{2}$$

$$A_{12} = 1.0; \quad A_{22} = \frac{1}{s_2 t_{0_K}}; \quad A_{32} = \frac{s_2}{2}$$

and $\qquad$ I = 1, 2, ..., m $\qquad$ [5] (m=1, 2, or 3)
$\qquad\qquad$ [6] J = 1, 2, 3
$\qquad\qquad$ [7] K = k = 1, 2, 3, ..., 15

Note: m is specified at either 1, 2, or 3.

v) $\qquad$ Print-out results in format:

$\qquad\qquad\qquad$ "t_{0_K} =x.xxxx , square step response"

t	Y11k	Y21k	Ym1k	θ11k	θ21k	θm

$$[8]\,\Delta t \begin{cases} 0 \\ .025 \\ .050 \\ .... \\ t_{Max} \end{cases}$$

$\qquad\qquad\qquad \pm$ x.xxx E0x $\quad \pm$ x.xxx E0x $\quad \pm$ x.xxx E0x $\qquad$ (same format)

---

[5] Number of forcing functions
[6] Types of loads
[7] Pulse time
[8] Fixed point, 3 decimals

"$\underline{t_{0_K}}$ =x.xxxx , ramp response"

t	Y12k	Y22k	Ym2k	θ12k	θ22k	θm2k
0	-	-	-	-	-	-
.						
.						
.						
t_{Max}	-	-	-	-	-	-

"$\underline{t_{0_K}}$ =x.xxxx , half sine response"

t	Y13k	Y23k	Ym3k	θ13k	θ23k	θm3k
0	-	-	-	-	-	-
.						
.						
.						
t_{Max}	-	-	-	-	-	-

<u>Input Data:</u>

$$s_1 = \text{x.xxxx E0x}$$
$$s_2 = \text{x.xxxx E0x}$$
$$t_{Max} = \text{x.xxx}$$
$$\Delta t = \text{x.xxx}$$
$$m = \text{x.} \quad (1, 2, \text{ or } 3)$$

$$\left.\begin{array}{l} \Gamma I1 \\ \Gamma I2 \\ LI1 \\ LI2 \end{array}\right\} \quad \text{format } \pm \text{x.xxxxE0x}, \qquad I = 1, 2, \dots, m$$

Z Axis Response

Input Data:

$$t_{Max} = 1.000$$
$$\Delta t = 0.025$$
$$m = 2$$
$$s_1 = 14.956 = 1.4956 \ E01$$
$$s_2 = 18.126 = 1.8126 \ E01$$

$$\Gamma_{i1,2} \begin{cases} \Gamma_{11} = 148.540 = +1.4854 \ E02 \\ \Gamma_{12} = 247.111 = +2.4711 E02 \\ \Gamma_{21} = -4.245 = -4.2450 E00 \\ \Gamma_{22} = 2.889 = +2.8890 E00 \end{cases}$$

$$L_{i1,2} \begin{cases} L_{11} = -328.048 = -3.2805 \ E02 \\ L_{12} = 326.624 = 3.2662 E02 \\ L_{21} = 1.246 = 1.2460 E00 \\ L_{22} = -1.071 = -1.0710 E00 \end{cases}$$

Y Axis Response

"Input Data"

$$t_{Max} = 0.300$$
$$\Delta t = 0.003$$
$$m = 3$$
$$s_1 = \sqrt{1.3027 \times 10^3} = 36.093 = 3.6093E01$$
$$s_2 = \sqrt{2.62663 \times 10^4} = 162.07 = 1.6207E02$$

$$\Gamma_{i1,2} \begin{cases} \Gamma_{11} = +1.0575E00 \\ \Gamma_{12} = +3.2560E-02 \\ \Gamma_{21} = +6.3453E00 \\ \Gamma_{22} = +1.9536E01 \\ \Gamma_{31} = +1.0575E00 \\ \Gamma_{32} = +3.2560E00 \end{cases}$$

$$L_{i1,2} \begin{cases} L_{11} = +1.7571E-03 \\ L_{12} = -1.5942E-03 \\ L_{21} = +1.0543E00 \\ L_{22} = -9.5646E-01 \\ L_{31} = +1.7571E-03 \\ \Gamma_{32} = -1.4071E-03 \end{cases}$$

Gimbal Z Axis Response

Input Data:

$$t_{Max} = 0.500 \qquad s_1 = 16.91 = 1.6910E01$$
$$\Delta t = 0.004 \qquad s_2 = 147.87 = 1.4787E02$$
$$m = 1$$

$$\Gamma \begin{cases} \Gamma_{11} = 1.07040 = 1.0704E00 \\ \Gamma_{12} = 1.05078 = 1.0508E00 \end{cases}$$

$$L \begin{cases} L_{11} = 0 \\ L_{12} = 0 \end{cases}$$

Appendix 5

Gimbal Ring and Arm System
Flexibility Coefficients

Using the notation of page 2, part B. I, flexibility coefficients for the gimbal ring may be obtained from the digital computer print-out of the finite element analysis of the ring.

For loading case XIII, we have:

$$\begin{cases} V_z = -.5 & kip \\ M_x = 10.25 & in\ kip \\ M_y = 54.0 & in\ kip \end{cases}$$

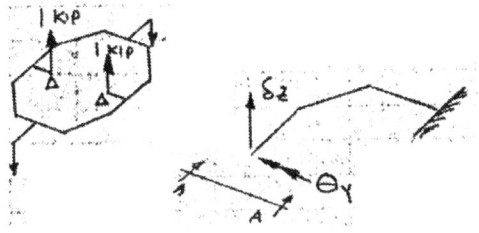

Referring to the unit analysis:

$$\delta_z = -.5(-2.06)10^{-3} - 8.26(10.25)10^{-6} - 3.27(54)10^{-5} = -0.814 \times 10^{-3}$$
$$\theta_y = -.5(2.834)10^{-5} + 7.69(10.25)10^{-7} - 1.47(54)10^{-6} = -5.733 \times 10^{-5}$$

View A-A

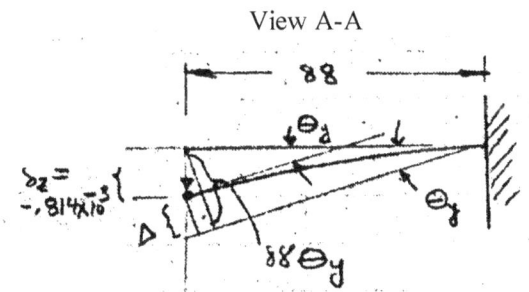

$$\Delta = 88|\theta_y| - |\delta_z| = 88(5.733 \times 10^{-5}) - 0.814 \times 10^{-3}$$
$$= 5.045 \times 10^{-3} - 0.814 \times 10^{-3}$$
$$= 4.231 \times 10^{-5}\ in.$$

Thus, $\qquad \Delta = 2.116 \times 10^{-6}\ in./lb.$ $\qquad$ (1/2 lb. at each load point)

$$\alpha_{22} = \alpha_{11} + \Delta$$

From Appendix 2, $\qquad \alpha_{22}$ = deflection δ_z due to unit load F_z
$$= 39.42 \times 10^{-6}\ in./lb.$$

Hence, $\begin{cases} \alpha_{11} = 39.42 \times 10^{-6} \equiv \alpha_{12} = \alpha_{21} \\ \alpha_{22} = 41.54 \times 10^{-6} \end{cases}$ $\qquad$ Flexibility coefficients

Appendix 6

Analysis of Coupling Effects
Of Arm and Gimbal

In order to analyze the coupling between the arm flexibility and the gimbal ring, the model of the system used in Part B.II will be studied more thoroughly.

Rotary inertia of the masses will be neglected and bending deformations will only be considered. The mass of the gondola will be concentrated at the center of the gimbal together with one half of the mass of the gimbal ring. The ring will be idealized as a mass-less spring. One fourth of the mass of the arm will be concentrated at the tip of the arm along with the mass of the forks and the remainder of the gimbal mass.

The stiffness of the arm and the gimbal ring are of different orders of magnitude. It follows that, if only transient forcing functions are allowed, the discrete (or "isolated") frequency method of analysis is applicable: i.e., that the system can be uncoupled and treated as a simple single degree of freedom system in order to obtain the "maximax" response of the arm.

This analysis is for the purpose of verifying the uncoupling of the system assumed tacitly to be possible in the analysis of Part A.

From page 3, Part B.I, the characteristic Eq. is:

$$p(s) = s^4(\alpha_{11}\alpha_{22} - \alpha_{12}\alpha_{21})m_1 m_2 + s^2(\alpha_{11}m_1 + \alpha_{22}m_2) + 1 \qquad (1)$$

We desire to investigate the effect of the variation in stiffness of the arm and gimbal ring structures on the response of the system.

Since the gimbal ring is considerably more rigid than the arm, we will study the effect of varying the gimbal rigidity on the response of the arm. In particular, we will be interested in the maximum transient loads that will act on the arm, or, somewhat more conveniently, the maximax displacements of mass #1.

Suppose that the gimbal ring is infinitely rigid. Then the flexibility coefficients in Eq. (1) are:

$$\alpha_{11} = \alpha_{22} = \alpha_{12} = \alpha_{21}$$

And, the characteristic Eq. reduces to

$$\alpha_{11}s^2(m_{Tot}) + 1 = 0$$

or, $$s^2 = \frac{-1}{\alpha_{11}m_{Tot}}$$

and $$\left\{ s_N^2 = \frac{1}{\alpha_{11}m_{Tot}} = \frac{1}{f} \right. \tag{2}$$

The transient response of the single degree of freedom system to the loading $F_z(t)$ will be:

$$u_1 = \frac{1}{f} \frac{\alpha_{11} f(s)}{(s^2 + s_N^2)} \tag{3}$$

Taking the inverse transform, we have:

$$Z_1 = \frac{1}{m_{Tot}} L^{-1} \left\{ \frac{f(s)}{s^2 + s_N^2} \right\} \qquad \text{for the single degree of freedom model.} \tag{4}$$

The solution for the transient response of the two degree of freedom system is, from Eq. (12), Chapter 2:

$$Z_1 = b_2 \sum_{j=1}^{2} A_{1j} L^{-1} \left\{ \frac{f(s)}{s^2 + s_j^2} \right\} \tag{5}$$

where:

$$b_2 = \frac{1}{\alpha_{11}\alpha_{22} - \alpha_{12}\alpha_{21}} \cdot \frac{1}{m_1 m_2}$$

$$A_{11} = \frac{\alpha_{12}}{s_2^2 - s_1^2}; \qquad A_{12} = -A_{11}$$

And:

$$\left\{ \begin{array}{l} s_1^2 = \frac{b_1}{2} - \sqrt{\left(\frac{b_1}{2}\right)^2 - b_2} \\ s_2^2 = \frac{b_1}{2} + \sqrt{\left(\frac{b_1}{2}\right)^2 - b_2} \end{array} \right.$$

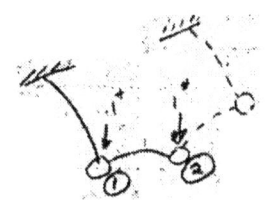

Where $\quad b_1 = \dfrac{\alpha_{11}m_1 + \alpha_{22}m_2}{\alpha_{11}\alpha_{22} - \alpha_{12}\alpha_{21}} \cdot \dfrac{1}{m_1 m_2}$

Now, $\alpha_{12} = \alpha_{21} \equiv \alpha_{11}$; $\qquad$ and let $\alpha_{22} = \alpha_{11} + \varepsilon\alpha_{11}$

Hence,
$$b_1 = \dfrac{\alpha_{11}m_1 + \alpha_{11}(1+\varepsilon)m_2}{\alpha_{11}(\alpha_{22} - \alpha_{11})} \cdot \dfrac{1}{m_1 m_2}$$
$$= \dfrac{m_1 + (1+\varepsilon)m_2}{\varepsilon\alpha_{11} \cdot m_1 m_2}$$

Also, $\quad b_2 = \dfrac{1}{\alpha_{11}(\alpha_{22} - \alpha_{11})m_1 m_2} = \dfrac{1}{\varepsilon\alpha_{11}^2 m_1 m_2}$

Thus, $\qquad s_2^2 - s_1^2 = 2\sqrt{\left(\dfrac{b_1}{2}\right)^2 - b_2} = \sqrt{b_1^2 - 4b_2}$

$$A_{11} = \dfrac{\alpha_{12}}{\sqrt{b_1^2 - 4b_2}} = -A_{12}$$

From Eq. (5):
$$Z_1 = b_2 A_{11} L^{-1}\left\{\dfrac{f(s)}{s^2 + s_1^2}\right\} + b_2 A_{12} L^{-1}\left\{\dfrac{f(s)}{s^2 + s_2^2}\right\}$$
$$= \dfrac{b_2 \alpha_{12}}{\sqrt{b_1^2 - 4b_2}} L^{-1}\left\{\dfrac{f(s)}{s^2 + s_1^2}\right\} - \dfrac{b_2 \alpha_{12}}{\sqrt{b_1^2 - 4b_2}} L^{-1}\left\{\dfrac{f(s)}{s^2 + s_2^2}\right\} \qquad (6)$$
$$= \dfrac{\alpha_{12}}{\sqrt{\left(\dfrac{b_1}{b_2}\right)^2 - \dfrac{4}{b_2}}}\left[L^{-1}\left\{\dfrac{f(s)}{s^2 + s_1^2}\right\} - L^{-1}\left\{\dfrac{f(s)}{s^2 + s_2^2}\right\}\right]$$

Now, $\dfrac{b_1}{b_2} = \dfrac{m_{Tot} + \varepsilon m_2}{\varepsilon\alpha_{11}m_1 m_2} \cdot \varepsilon\alpha_{11}^2 m_1 m_2 = (m_{Tot} + \varepsilon m_2)\alpha_{11}$

$$\frac{4}{b_2} = 4\varepsilon\alpha_{11}^2 m_1 m_2$$

$$\therefore \sqrt{\left(\frac{b_1}{b_2}\right)^2 - \frac{4}{b_2}} = \sqrt{m_{Tot}^2 \alpha_{11}^2 + 2m_{Tot}\alpha_{11}^2 m_2 \varepsilon + \varepsilon^2 m_2^2 \alpha_{11}^2 - 4\varepsilon\alpha_{11}^2 m_1 m_2}$$

$$= \sqrt{m_{Tot}^2 \alpha_{11}^2 + 2\varepsilon\alpha_{11}^2 m_1 m_2 + 2\varepsilon\alpha_{11}^2 m_2^2 + \varepsilon^2 \alpha_{11}^2 m_2^2 - 4\varepsilon\alpha_{11}^2 m_1 m_2}$$

$$= \sqrt{m_{Tot}^2 \alpha_{11}^2 - 2\varepsilon\alpha_{11}^2 m_1 m_2 + \alpha_{11}^2 m_2^2 (2\varepsilon + \varepsilon^2)}$$

By hypothesis, $\varepsilon \ll 1$,

Hence $\quad \sqrt{\left(\frac{b_1}{b_2}\right)^2 - \frac{4}{b_2}} \cong \dot{m}_{Tot}\alpha_{11}$

And Eq. (6) reduces to:

$$Z_1 = \frac{\alpha_{11}}{m_{Tot}\alpha_{11}}\left[L^{-1}\left\{\frac{f(s)}{s^2 + s_1^2}\right\} - L^{-1}\left\{\frac{f(s)}{s^2 + s_2^2}\right\}\right]$$

$$= \frac{1}{m_{Tot}1}\left[L^{-1}\left\{\frac{f(s)}{s^2 + s_1^2}\right\} - L^{-1}\left\{\frac{f(s)}{s^2 + s_2^2}\right\}\right] \qquad (7)$$

Now, $b_1 = \dfrac{m_{Tot} + \varepsilon m_2}{\varepsilon\alpha_{11}m_1 m_2}$

And, $\quad \sqrt{\left(\frac{b_1}{2}\right)^2 - b_2} = \frac{b_2}{2}\sqrt{\left(\frac{b_1}{b_2}\right)^2 - \frac{4}{b_2}}$

$$\cong \frac{1}{2\varepsilon\alpha_{11}^2 m_1 m_2} \cdot m_{Tot}\alpha_{11} = \frac{m_{Tot}}{2\varepsilon\alpha_{11}m_1 m_2}$$

Hence, $\quad s_1^2 \cong \dfrac{m_{Tot} + \varepsilon m_2}{2\varepsilon\alpha_{11}m_1 m_2} - \dfrac{m_{Tot}}{2\varepsilon\alpha_{11}m_1 m_2} = \dfrac{\varepsilon m_2}{2\varepsilon\alpha_{11}m_1 m_2} = \dfrac{1}{2\alpha_{11}m_1}$

$$s_2^2 \cong \frac{m_{Tot} + \varepsilon m_2}{2\varepsilon\alpha_{11}m_1 m_2} + \frac{m_{Tot}}{2\varepsilon\alpha_{11}m_1 m_2} \cong \frac{m_{Tot}}{\varepsilon\alpha_{11}m_1 m_2}$$

$$\text{since } \varepsilon \ll 1$$

$$\boxed{\therefore s_2^2 \gg s_1^2} \quad \text{(as originally hypothesized)}$$

Now,

$$L^{-1}\left\{\frac{f(s)}{(s^2+s_j^2)}\right\} = \frac{1}{s_j}\int_0^t F_z(\beta)\sin s_j(t-\beta)d\beta,$$

using the convolution theorem.

Restricting our investigations to the type of functions already used (harmonic forcing functions can be neglected, since it has been shown in Part C that they have, at maximum, extremely low frequencies relative to the fundamental system frequencies), the inverse transforms have the form:

$$L^{-1}\left\{\frac{f(s)}{(s^2+s_j^2)}\right\} = F_{z_{Max}} \cdot \frac{f(\sin s_j t, \cos s_j t)}{(s_j)^m} \tag{8}$$

where $m \geq 1$

Since $s_2^2 \gg s_1^2$, and $s_2 \gg s_1$, in view of Eq. (8), where the function $f(\sin s_j t, \cos s_j t) \leq 2$, the 2nd term in Eq. (7) is negligible. Hence, Eq. (7) reduces to Eq. (4), page 4. In short, the response of mass 1 can be accurately predicted assuming that the gimbal ring is infinitely rigid as long as

$$\varepsilon \ll 1$$

or, equivalently, $s_2^2 \gg s_1^2$

Using the data from Appendix 5, we have:

$$\alpha_{11} = 39.42\times10^{-6}$$
$$\alpha_{12} = \alpha_{21} = \alpha_{11}$$
$$\alpha_{22} = \alpha_{11}(1+\varepsilon) = 39.42\left(1+\frac{2.12}{39.42}\right) = 39.42(1+.0538)$$
$$= 41.54\times10^{-6}$$

Now, $\varepsilon = .0538$ is $\ll 1$

Hence, uncoupling the system is justifiable.

131

Also, from page 9, Part B.II:

$$\begin{cases} s_1^I = \sqrt{286.0} = 16.91 \quad rad\,/\sec;\ f_1 = \dfrac{s_1}{2\pi} = 2.69 \quad cps \\[2mm] s_2^I = \sqrt{21{,}866} = 147.87 \quad rad\,/\sec;\ f_2 = \dfrac{s_2}{2\pi} = 23.53 \quad cps \end{cases}$$

Note that $s_2 \gg s_1$, as required.

These frequencies were derived neglecting rotary inertia. As a result, they should be somewhat higher than those which would result if it were considered. The fundamental frequency can be compared to that obtained in Chapter 1, page 51, wherein rotary inertia was considered and the system was analyzed in the uncoupled state. There,

$$s_1^{II} = 14.96 \quad rad\,/\sec, \quad f_1 = 2.38 \quad cps$$

Also, from Eq. (2), page 3 of this appendix, the fundamental frequency was derived ignoring rotary inertia and the gimbal ring flexibility. There, we obtained

$$s_N^2 = \frac{1}{\alpha_{11} m_{Tot}} = \frac{10^6}{39.42 \times 87.45} = 290.08$$

or, $\quad s_N = \sqrt{290.08} = 17.03 \quad rad\,/\sec, \quad f_1 = 2.71 \quad cps$

Since s_1^I and s_N differ only slightly, the gimbal ring is observed to have negligible effect on the fundamental frequency. Hence, the difference, between s_1^I and s_1^{II} is not due to the uncoupling of the system, but instead, to the inclusion of rotary inertia!

It may be concluded that the system can be uncoupled and studied separately without disadvantage as far as response of mass 1 is concerned.

Appendix 7
Photographs

Photograph of the MSC Centrifuge

Figure 7.1

www.ingramcontent.com/pod-product-compliance
Lightning Source LLC
Chambersburg PA
CBHW081127170526
45165CB00008B/2581